PACIFIC PROFILES

VOLUME 16
Allied Bombers: B-17 Flying Fortress series
Australia, New Guinea and the Solomons
1942–1944

MICHAEL JOHN CLARINGBOULD

Avonmore Books

Pacific Profiles Volume 16

Allied Bombers: B-17 Flying Fortress series Australia, New Guinea and the Solomons 1942–1945

Michael John Claringbould

ISBN: 9780975642351

First published 2024 by Avonmore Books
Avonmore Books
PO Box 217
Kent Town
South Australia 5071
Australia

Phone: (61 8) 8431 9780
avonmorebooks.com.au

A catalogue record for this book is available from the National Library of Australia

Cover design & layout by Diane Bricknell

© 2024 Avonmore Books.

No part of this book may be reproduced or transmitted in any form or by any means, electronic or mechanical, including photocopying or recording, or by any information storage and retrieval system, without permission in writing from the publisher.

Front Cover: This selection of four Fortresses showcases the diverse nature of camouflage schemes which appeared in all three theatres, as illustrated in Profiles 1, 80, 103 and 20. The top Fortress was painted in the five-tone Hawaii Air Depot Scheme, whilst the B-17F underneath was distinguishable by a particularly unusual green camouflage pattern on the fin. Pretty Baby is illustrated as it served with the 40th Troop Carrier Squadron as squadron number 41. The lowest Fortress retained its British serial from when it was originally earmarked for RAF Coastal Command as a Fortress Mark IIA. It retained its RAF markings when subsequently reassigned to the USAAF, arriving at Mareeba in August 1942.

Back Cover: An 11th BG B-17E drops supplies to a forward US Marine position along the northern Guadalcanal coast past the Matanikau River in late 1942. Such air-drops included medical supplies, ammunition, rations and at times even water.

Contents

About the Author 5
Map 6
Glossary & Abbreviations 7
Introduction 9
Chapter 1   Markings and Technical Notes 11
Chapter 2   Bombardment Groups 19
Chapter 3   14th Reconnaissance Squadron 25
Chapter 4   23rd Bombardment Squadron 29
Chapter 5   26th Bombardment Squadron 33
Chapter 6   28th Bombardment Squadron 39
Chapter 7   30th Bombardment Squadron 43
Chapter 8   31st Bombardment Squadron 47
Chapter 9   40th Reconnaissance Squadron and 435th Bombardment Squadron 51
Chapter 10   42nd Bombardment Squadron 57
Chapter 11   63rd Bombardment Squadron 63
Chapter 12   64th Bombardment Squadron 71
Chapter 13   65th Bombardment Squadron 75
Chapter 14   72nd Bombardment Squadron 79
Chapter 15   93rd Bombardment Squadron 83
Chapter 16   98th Bombardment Squadron 89
Chapter 17   394th Bombardment Squadron 93
Chapter 18   403rd Bombardment Squadron 99
Chapter 19   8th Photo Reconnaissance Squadron 103
Chapter 20   431st Bombardment Squadron 107
Chapter 21   2nd Provisional Bombardment Squadron 111
Chapter 22   Commanders' Transports 115
Chapter 23   Armed Transports 119
Sources & Acknowledgments 123
Index of Names 124

The author in 2004 with 14th RS B-17E 41-2446 which force-landed in New Guinea's Agaiambo Swamp on 22 February 1942.

About the Author

Michael Claringbould – Author & Illustrator

Michael spent his formative years in Papua New Guinea in the 1960s, during which he became fascinated by the many WWII aircraft wrecks which still lie around the country. Michael has served widely overseas as an Australian diplomat, including in South East Asia and throughout the South Pacific where he had the fortune to return to Papua New Guinea for three years commencing in 2003. Michael has authored and illustrated many books on Pacific War aviation, including over 30 titles for Avonmore Books. His history of the Tainan Naval Air Group in New Guinea, *Eagles of the Southern Sky*, received worldwide acclaim as the first English-language history of a Japanese fighter unit, and was translated into Japanese. An executive member of Pacific Air War History Associates, Michael holds a pilot license and PG4 paraglider rating. He continues to develop his skills as a digital 3D aviation artist, using 3DS MAX, Vray and Photoshop to attain markings accuracy.

Other volumes in this series:

Pacific Profiles Volume One Japanese Army Fighters New Guinea & the Solomons 1942–1944 (2020)

Pacific Profiles Volume Two Japanese Army Bomber & Other Units, New Guinea and the Solomons 1942–44 (2020)

Pacific Profiles Volume Three Allied Medium Bombers, A20 Series, South West Pacific 1942–44 (2020)

Pacific Profiles Volume Four Allied Fighters: Vought F4U Corsair Series Solomons Theatre 1943–1944 (2021)

Pacific Profiles Volume Five Japanese Navy Zero Fighters (land-based) New Guinea and the Solomons 1942–1944 (2021)

Pacific Profiles Volume Six Allied Fighters: Bell P-39 & P-400 Airacobra South & Southwest Pacific 1942–1944 (2022)

Pacific Profiles Volume Seven Allied Transports: Douglas C-47 series South & Southwest Pacific 1942–1945 (2022)

Pacific Profiles Volume Eight IJN Floatplanes in the South Pacific 1942–1944 (2022)

Pacific Profiles Volume Nine Allied Fighters: P-38 series South & Southwest Pacific 1942–1944 (2022)

Pacific Profiles Volume 10: Allied Fighters: P-47D Thunderbolt series Southwest Pacific 1943–1945 (2023)

Pacific Profiles Volume 11: Allied Fighters: USAAF P-40 Warhawk series South and Southwest Pacific 1942–1945 (2023)

Pacific Profiles Volume 12: Allied Fighters: P-51 & F-6 Mustang series New Guinea and the Philippines 1944–1945 (2023)

Pacific Profiles Volume 13: IJN Bombers, Transports, Flying Boats & Miscellaneous Types South Pacific 1942–1944 (2024)

Pacific Profiles Volume 14: Allied Bombers: B-25 Mitchell series Australia, New Guinea and the Solomons 1942–1945 (2024)

Pacific Profiles Volume 15: Allied Bombers B-26 Marauder series Australia, New Guinea and the Solomons 1942–1944 (2024)

Map

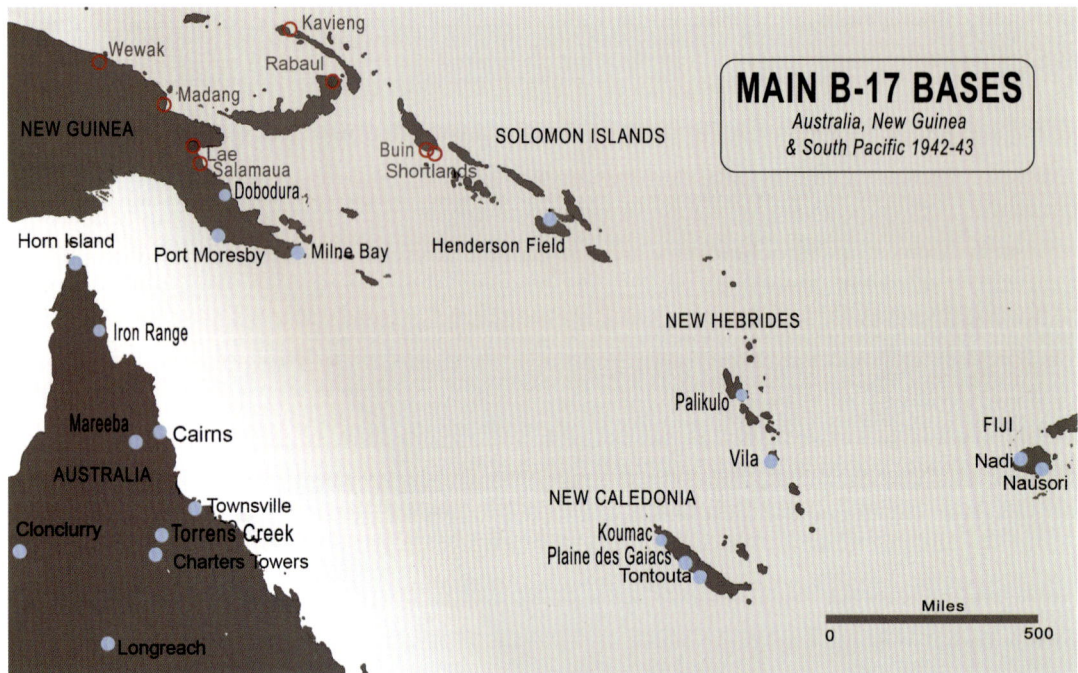

A map of the main bases used by B-17s throughout Australia, New Guinea and the South Pacific during 1942-43, shown in blue. Key Japanese bases and targets are shown in red.

In the early months of 1942, B-17s were based at inland Australian locations such as Cloncurry and Longreach. This avoided coastal airfields where the large bombers, almost impossible to camouflage, were vulnerable to attack. To strike key Japanese targets in New Guinea the bombers usually refuelled at Seven-Mile 'drome, Port Moresby, on the outgoing or return leg. Such tactics successfully avoided the destruction of the Fortresses during Japanese bomber raids, but even with the development of fields further north at Mareeba and Iron Range the enormous distances flown put great strain on the fleet. By late 1942 the 19th BG inventory of B-17s was all but exhausted, and the group returned to the US. By 1943 the replacement 43rd BG enjoyed the relative luxury of operating from New Guinea bases such as Port Moresby.

Likewise in the South Pacific the 11th BG B-17s were initially distributed throughout the New Hebrides, New Caledonia and Fiji. Aircraft staged through Palikulo in Espirito Santo to attack targets in the Solomons. Operations from Henderson Field were initially limited by continuous Japanese air attacks on the location, and also by the availability of fuel. However, from late 1942 B-17s were increasingly based there, greatly aided by the construction of a second bomber strip called Carney Field.

Glossary & Abbreviations

Note: Japanese terms are in italics.

ASV	Air to Surface Vessel (radar)
BG	Bombardment Group
BS	Bombardment Squadron
FPO2c	Flying Petty Officer second class
HAD	Hawaii Air Depot
Hiko Sentai	A Japanese Army Air Force flying regiment
Kokutai	A Japanese naval air group
MIA	Missing in Action
POW	Prisoner of War
PRS	Photo Reconnaissance Squadron
RAAF	Royal Australian Air Force
RAF	Royal Air Force
RNZAF	Royal New Zealand Air Force
RS	Reconnaissance Squadron
SOPAC	South Pacific Area
SS	Single screw steamship
SWPA	South West Pacific Area
TCG	Troop Carrier Group
TCS	Troop Carrier Squadron
The Slot	A colloquial term, used by the Allies, referring to the NW/SE geographic ocean route which lay within the Solomon Islands.
US	United States
USAAC	United States Army Air Corps
USAAF	United States Army Air Force
USMC	United States Marine Corps
USN	United States Navy
USS	United States Ship
VIP	Very Important Person

A timeline showing the deployment of B-17 units in Australia, New Guinea and the South Pacific during 1942-44.

Introduction

This volume details the markings of the B-17 Flying Fortress in the SWPA, SOPAC and Australian theatres. A varied mixture of misfortune and achievement accompanied the type whose markings are much complicated by the fact that nearly every Fortress in these theatres served at least two or more squadrons. Many spanned service lives of over three years.

The Boeing B-17E was the state-of-the-art heavy bomber in the USAAC inventory when the Pacific War began and was already serving in the Philippines at the war's outbreak. Eventually both the "E" and "F" models would serve throughout the Pacific until late 1944 with two provisional and eighteen mainstay USAAF bombardment squadrons as well as a reconnaissance unit. Others served as converted transports and commanders' aircraft.

The first appearance of a Fortress in the Australia-New Guinea area pre-dated hostilities, when 26 "D" models ferried themselves in late 1941 to the Philippines via Hawaii, Port Moresby and Darwin. Until this landmark flight, long-range deliveries to the Philippines negotiated an elongated ferry route through Africa, India and the NEI. This first Pacific delivery incorporated similar navigational challenges employed by aviatrix Amelia Earhart and navigator Fred Noonan some five years previous. No Fortress crew had previously attempted such long-distances before, made by dead-reckoning and astral fixes.

The inaugural Pacific route went through Hawaii prior to the longest sector from Wake Island to Port Moresby. The fourteen-hour journey took the bombers over the Japanese mandated territory of the Caroline Islands. Political tensions ensured no diplomatic clearance was sought from the Japanese government. As such, the sector was flown at night at 25,000 feet, assessed as beyond the range of Japanese fighters, with oxygen and no running lights. All defensive guns were loaded in anticipation of enemy interception, even though no state of war existed. Once past the Caroline Islands, the bombers descended to lower altitudes. The next sector was to Darwin, a flight of six hours. Remarkably, no aircraft were lost, which was not the case for subsequent wartime delivery flights across the Pacific where a mixture of Mitchells, Marauders and Liberators disappeared on almost every ferry flight.

Whilst this pre-war ferry journey marked the first appearance of Fortresses in New Guinea, in the SOPAC theatre the type's first appearance commenced on 6 January 1942, when three 23rd BS B-17Es (serials 41-2429, -2432 and -2433) departed Hawaii to assess the feasibility of conducting sea searches from island land bases including Nadi on Fiji. The reconnaissance mission was designed to also support the carrier USS *Enterprise* and to protect a USN convoy reinforcing Samoa at the time.

Shortly after the Battle of Midway, the role of Fortress groups in the Pacific was envisaged as a "Mobile Force", to provide armed searches to locate enemy ships. It was in this capacity that the 11th BG was first sent to the SOPAC prior to the Guadalcanal campaign to support USN operations. The unique *ad hoc* 14th RS was similarly dispatched to the SWPA to also operate under USN command, in order to cover Vice Admiral Wilson Brown's Task Force 11 centred on the USS *Lexington* as it headed for Rabaul. The unit was commandeered and assembled from Hawaii's Seventh Air Force at short notice, with its fundamental objective reverting to the protection of Australia's shipping supply lines.

Despite inflated claims, in reality the Fortress' record against shipping from higher altitudes was lacklustre, for the Japanese learned quickly to alter course as soon as bombs were released. As an alternative, Major William Benn, commanding the 63rd BS, was tasked to develop a low-level skip-bombing technique for the four-engine bomber. Training was finessed against shipwreck SS *Pruth* off Port Moresby. Despite optimism the tactic proved far more difficult to execute in practice than anticipated, and practical opportunities to sink ships were limited.

Neither were field modifications lacking on the type. In the SOPAC theatre the 11th BG fitted several B-17Es with SCR-521 long-wave Air to Surface Vessel (ASV) search radar. This initiative also incurred limitations, not the least being the degree to which tropical heat and humidity played havoc with valves and circuitry. Engineers at Espiritu Santo replaced the early Sperry turret system in December 1942 by installing twin 0.50-inch calibre machine guns on a flexible mount facing downwards through the circular fuselage opening. Engineers in Hawaii later fitted twin 0.50-inch guns in the nose to deter frontal attacks.

The SOPAC Fortresses carved a separate niche in history when they frequently battled it out with Mavis flying boats, such encounters being more frequent than acknowledged. Then, from mid-1943 onwards B-17s battled it out with J1N1 night-fighters, both over Rabaul and Bougainville. Pacific Fortresses performed hundreds of long-range reconnaissance and photographic flights too, in addition to hauling disassembled artillery pieces over the Owen Stanley mountains and dropping water, supplies and medicines to exhausted American troops on Guadalcanal. The bombers also flew VIP flights and rotated personnel on leave.

By the end of 1943 the Fortress had all but been replaced by the B-24 Liberator, however, workshops stripped and converted many into transport work horses, a far cry from the original concept of a "Mobile Force".

There is no better way to illustrate the extended service of the type in the Pacific than by the legacy of B-17E 41-2464 *Queenie*, the last Fortress lost in the theatre (see Profile 107). It served numerous units in differing capacities before seeing out its final days as a transport. Finally on 8 July 1944, the veteran disappeared between Nadzab and Tadji, costing the lives of nineteen aboard.

In New Guinea even today the Fortress story continues. As recently as 2023, the wreckage of B-17F 41-24427 was located in a mountainous part of New Britain. It was lost with Captain Robert Williams' crew on 15 September 1942 when heading to Rabaul at night.

Michael Claringbould
Canberra August 2024

CHAPTER 1
Markings and Technical Notes

Welcome to the markings of Pacific and Australian-based B-17E and B-17F model Flying Fortresses. This volume details key markings specifications and modifications to the type as unique to Pacific operations, and covers many Fortresses not illustrated to date. The demands of the SWPA and SOPAC theatres produced several markings eccentricities, including bombers originally slated for Europe that were diverted for Pacific use, and hence entered the theatre with RAF camouflage schemes.

Every *Pacific Profiles* volume dispels myths and corrects errors which have become entrenched over the years. Whilst accurate profiles of European theatre B-17s are ubiquitous, accurate depictions of Pacific Fortresses are mostly lacking, especially from the SOPAC theatre. Much illustrative work to date suffers from confused research, and as a result the type in Pacific service is poorly illustrated. In the worst case, some publications have even replicated profiles of non-existent airframes!

The quintessential example of this mythology is the markings and paint scheme of *Blue Goose* (see Profile 70). Technical documentation exists for this scheme, as opposed to the speculation and unsubstantiated theories which have produced a suite of fiction.

No less than a dozen B-17E/Fs went on to serve as transports in both the SOPAC and SWPA theatres, all of which have rich histories, and some are illustrated here for the first time. So are commanders' Fortresses and those assigned to provisional squadrons. This lost legion of refurbished Fortresses played a substantive role in the Pacific War and should not be sidelined.

It is pleasing to define and illustrate the "Hawaii Air Depot" (HAD) scheme, a five-tone *ad hoc* scheme applied at Hawaii immediately post-Pearl Harbor. Every B-17E which arrived in the immediate aftermath of the infamous 7 December 1941 attack was commandeered by the Hawaii USN command. All arrived in Hawaii painted in Boeing's Olive Drab/Neutral Grey scheme, however, they were immediately utilised for patrol duties by Brigadier General Jacob Rudolph, the commander of the 18th Bombardment Wing. Rudolph decided that, regardless of whether they were conducting patrols or just being used for training, the bombers needed more effective camouflage.

A subsequent *ad hoc* scheme evolved, applied by the Air Depot at Hickam Air Force Base. Four separate colours were sprayed over the extant olive drab using USN paint stocks on hand. No two schemes were alike; they differed in both the applied patterns and even colour shade, and no technical orders were issued for implementation. The only common feature to all airframes is that the fin area containing the yellow USAAC serial stencil was left extant in order to retain the original serial. Of the four colours, the most distinctive was USN blue grey, with the other three approximate replications of light sand, rust brown and foliage green. The resulting five-colour scheme was unique to the USAAC at this time and was never repeated.

Nonetheless several Hawaii-based B-17Es escaped the scheme, however, records and photos confirm that at least two dozen B-17Es underwent the spray gun at Hickam's workshops, although the final number is likely closer to 30. These two dozen confirmed recipients are B-17E serials 41-2397, -2404, -2408, -2409, -2413, -2416, -2417, -2420, -2421, -2426, -2428, -2429, -2430, -2432, -2433, -2434, -2437, -2442, -2444, -2445 and -2467 plus also 41-9059, -9156 and -9209. The HAD scheme is illustrated at the top of Figure 1 on page 14.

Among the escapees who originally avoided the paint gun was a trio which left Hawaii on 16 January 1942 to support the USS *Enterprise* and protect a convoy *en route* to Samoa. These then maintained reconnaissance along the Pacific delivery route to Australia. All three were assigned to the 23rd BS, being 41-2429, -2432 and -2433. These subsequently received the HAD scheme upon return to Hawaii.

Brigadier General Rudolph's ordained camouflage scheme incorporated red and white striped rudders, applied in equidistant bands of seven red and six white stripes. The purpose of this garish addition was derived from the fact that during the Pearl Harbor attack several B-17Es had been fired at by nervous "friendly gunners". It was hoped that the bright rudders would prevent similar occurrence. These rudders differed to the official USAAC pre-war marking insofar as they lacked a blue leading vertical band.

It is equally satisfying to present B-17Es which appeared in RAF camouflage, reclaimed by the USAAC from a British consignment. At least four B-17Es entered Pacific service in RAF Temperate Sea and Sky camouflage, as shown in the middle aircraft in Figure 1. These four were commandeered by the USAAC from a batch of sixteen delivered to Cheyenne for modification to RAF specifications: serials 41-2609, 41-9234, -9235 and -9244. Of these, 41-9234 was the only one to retain its RAF Air Ministry serial FL481, still faintly visible on the wreck even today (see Profile 20). At least one more B-17 entered SOPAC service with a different RAF scheme (see Profile 70).

B-17s served not only as bombers in the Pacific but were also given more mundane tasks. Here a 64th BS B-17E is loaded with fresh produce at 30-Mile airfield (Rorona) around mid-1943, which had been grown in nearby gardens. The aircraft is B-17E 41-2648 Stud Duck, which is illustrated in Profile 81 when later serving with the 403rd BS. Note the radar aerials on the side and top of fuselage.

Taken at Seven-Mile 'drome, Port Moresby, on 2 November 1941, this natural metal finish B-17D was one of the first Fortresses to appear in the SWPA.

A field-modified 43rd BG B-17E at Port Moresby with twin 0.50-inch calibre machine guns installed in the nose. Several SOPAC Fortresses had the same modification undertaken in Hawaii before deployment.

Figure 1 B-17E/F Markings & Technical

A. The USAAF serial was not applied to the first batches of B-17Es leaving the Boeing factory at Seattle. Serial stencils were subsequently applied in the field as circumstances permitted, explaining why differing sizes and styles appear in different timeframes. Whilst B-17Es leaving the factory from February 1942 had factory-applied serials, it is possible this practice commenced as early as late November 1941.

B. Brigadier General Rudolph's red and white striped rudders, applied on the HAD camouflaged airframes.

C. "US ARMY" was painted in dark blue (USAAC stencil format) across the undersurface of both wings, left to right, and facing forward.

D. All B-17Es within the serial range 41-2393 to 41-2504 left the Boeing factory equipped with a Sperry remote turret. The system incorporated twin 0.50-inch calibre machine guns sighted via an observation station positioned directly behind the turret, the operator facing aft in a prone position using a periscope. The unconventional system proved unsatisfactory, first as gunners had difficulty in acquiring fast-moving targets, and second as facing aft could induce nausea. Accordingly, many Pacific crews replaced the remote turret system with a manned turret where available, or twin 0.50-inch guns on a flexible mount facing downwards through the circular fuselage opening.

E. The final belly turret installed was the Bendix ball turret, on B-17Es in the serial ranges 41-2505 to 41-2669 and 41-9011 to 41-9245 inclusive. The gun sight was mounted between the guns (this same equipment was also installed in the B-25B through to the first B-25G model, and also the B-24D). Bendix-equipped B-17s were oft photographed with the guns at vertical azimuth as they had to be stowed in this position to allow opening of the turret's access hatch inside the fuselage. Note that, unlike other installations in other types, the turret was not retractable in the B-17.

H. On RAF camouflage schemes the previous RAF roundel was painted over and later replaced with a US insignia.

J, K, & L. The 11th BG had several B-17Es fitted with SCR-521 long-wave Air to Surface Vessel (ASV) search radar, installed under the port wing (L), with a nose antenna (J) and extra dorsal aerial (K). Ground crews gave the colloquial name of "towel racks" to the wing-mounted antennas.

M. On 6 May 1942 a USAAF markings directive ordered all red/white striped rudders and the red centers of US national insignia be removed to avoid confusion with Japanese markings.

N. Despite hundreds of electrical and airframe modifications resulting in the B-17F model, its most distinctive feature was the revised Perspex nose. This can sometimes be a source of identity confusion, as several "E" models were fitted with replacement "F" model Perspex noses due to combat damage, and vice versa as illustrated in several profiles.

O. All B-17Fs which served in the Pacific lay in the serial range 41-24353 to 41-24621 inclusive, meaning "F" models are readily identifiable with six-digit stencils.

Figure 2
B-17E/F Series Stencils

P

U.S. ARMY MODEL B-17E
AIR CORPS SERIAL NO. 41-2504
CREW WEIGHT 1200 LBS

Q

NO PUSH

HOIST

NO STEP

BATTERY LOCATION HERE

R

100 OCT GALS

S

1234567 89

123456789

123456789

Figure 2 B-17E/F series Stencils

A host of stencil markings attended early "E" and "F" model Fortresses, several of which are illustrated here.

P. This manufacturer's stencil was applied at the Boeing factory in Seattle, conforming to USAAC requirements for airframe identification.

Q. A series of self-explanatory stencils as applied to different parts of the airframe.

R. Common stencils applied to the red fuel caps on the wing top surface.

S. A variety of serial number styles as applied in the field and/or the factory. Some were painted and others applied with stencils. Note the unique shape of the numerals "2" and "3" in the middle row.

A B-17E fitted with Air to Surface Vessel (ASV) radar at Espiritu Santo in December 1942. An under wing "towel rack" antenna is circled.

Taken at Cloncurry, Queensland, shortly after arriving in Australia in February 1942, 41-2416 was among the twelve B-17Es of the Carmichael Detachment. Note the Hawaii Air Depot scheme and red and white striped rudder. The original serial number had been mistakenly painted over in Hawaii and the last four digits of its serial have just been re-applied (see Profile 1).

CHAPTER 2
Bombardment Groups

This chapter summarises the histories of the four bombardment groups that served in the SWPA and SOPAC theatres: the 5th, 11th, 19th and 43rd Bombardment Groups. Subsequent chapters detail the component squadrons of these groups.

5th Bombardment Group

In December 1941 the 5th BG was operating B-17s and B-18s, and in February 1942 it was assigned to the Seventh Air Force. Its aircraft primarily undertook search and patrol missions from Hawaii in the months following Pearl Harbor.

The 5th BG commander was Lieutenant Colonel Edwin Bobzien from April 1941 until January 1942, then Colonel Arthur Meehan until October 1942, then Lieutenant Colonel Brooke Allen until August 1943 (who was promoted full colonel in February 1943) and then Lieutenant Colonel Marion Unruh until the end of 1943. The group operated four Fortress squadrons in the South Pacific: the 23rd, 31st, 72nd and 394th BS.

The 72nd BS was the first squadron to deploy to the SOPAC on 15 September 1942 when eight B-17Es headed for Espiritu Santo, with key engineering personnel ferried from Oahu in LB-30 Liberators. Meanwhile the group's other Fortresses continued to operate from various Hawaiian airfields, with the 31st BS, for example, using Kipapa Field on Oahu from 23 May 1942.

In order to facilitate the 5th BG's arrival in theatre, the commander of the 11th BG, Colonel Laverne "Blondie" Saunders, permitted the 5th BG airmen to use to use his group's B-17s. As but one example, the 72nd BS was fully incorporated into the 11th BG from 7 October 1942 for a period of several weeks. To underline the transitory nature of Fortress assignment in the SOPAC at this stage of the war, at the end of February 1943 this squadron's entire B-17 inventory was transferred to the 31st and 23rd BS when it commenced transition to Liberators.

The 394th BS experienced the most abnormal B-17 deployment, however, staying in Hawaii throughout 1942 from where it continued patrols and crew training. In early November 1942 it moved to Bellows Field equipped with seven ex-19th BG B-17E veterans which had been ferried back from Australia, then refurbished in Hawaii, before later moving to the SOPAC. On 4 January 1943 the 5th BG was transferred from the Seventh to the Thirteenth Air Force.

In the SOPAC theatre the considerable movement and interchange of the group's B-17s between units was much driven by engine reliability issues - many SOPAC Fortresses logged considerable "three-engine" time as a result of engine failures, and many replacement engines were those overhauled at Brisbane's Eagle Farm workshop. At the end of February 1943, the 11th BG transferred its entire remaining B-17 inventory to the 5th BG. The 23rd, 31st and 72nd BS commenced conversion to the B-24 Liberator from mid-1943 onwards, with the 394th BS following in early 1944.

11th Bombardment Group

First created as the 11th Observation Group on 1 October 1933, it became the 11th Bombardment Group (Medium) on 1 February 1940 at Hickam Field, comprising the 14th, 26th and 42nd BS along with the attached 50th Reconnaissance Squadron. Shortly after the Battle of Midway, the group was redesignated a "Mobile Force" by Washington, tasked to provide armed search aircraft to find Japanese fleets. In doing so it was expected that the group's bombers would have sufficient firepower to protect themselves from Japanese fighters. It was in this capacity that the 11th BG was sent to the South Pacific at the start of the Guadalcanal campaign to support USN operations. Colonel Laverne "Blondie" Saunders took the group's B-17s to war and commanded the unit until December 1942 when he returned to Hawaii and was replaced by Colonel Frank Everest.

Colonel Laverne "Blondie" Saunders was the 11th BG commander and a key figure in SOPAC B-17 operations. He survived the ditching of B-17F 41-24531 in the Solomons on 18 November 1942 after it was attacked by Zeros.

The 11 BG's first B-17Es were spread throughout the New Hebrides, Fiji and New Caledonia largely because each base only had minimal space and infrastructure. In late July 1942 the 98th BS arrived at Koumac, the 42nd BS at Plaine des Gaiacs, the 26th BS at Port Vila and the 431st BS at Nadi. The USN directed the group to hit Tulagi and Guadalcanal from 31 July to 6 August 1942 as a precursor to the US Marine invasion of Guadalcanal. Since the closest airfield, Palikulo on Espiritu Santo, could not accommodate B-17s until 1 August, the group opened these missions from Efate instead.

As the Solomons campaign unfolded, the 11th BG flew extended searches daily during which they often encountered four-engine Mavis flying boats whose 20mm cannon were out-ranged by the B-17s' own 0.50-inch machine guns. This meant the Fortresses could engage on their own terms, and by the end of September 1942 21 engagements with these flying boats had occurred, resulting in claims of five destroyed and seven damaged.

A quirk in the 11th BG's SOPAC deployment occurred in late 1942 when a dozen B-17s arrived at Port Moresby for a week's temporary duty on 29 December 1942. They were sent there as the base was closer to Rabaul as opposed to Guadalcanal, which was not only further but was also experiencing fuel shortages at the time. Once the shortage was resolved, Vice Admiral "Bull" Halsey requested his bombers be returned, and in the event the detachment only flew two missions in the SWPA theatre.

The ongoing poor state of Henderson Field remained an ongoing hindrance throughout 1942. Furthermore, the USN contended that the Fortresses' high altitude bomb delivery was ineffective against shipping (as proven to be the case from post-war surveys). Nonetheless the

USN had other ways to use the 11th BG: "Quackenbush's Gypsies" became the moniker for Lieutenant Commander Robert Quackenbush's USN photographic personnel who regularly flew aboard B-17s on reconnaissance missions.

A combination of tropical weather and a lack of navigation aids cost more B-17s than enemy fighters, and early operations were further hampered by crude support facilities. Lack of spare parts plagued the early deployments too, and the list of critical components needing replacement was incessant. These included *inter alia* broken turret doors and supercharger components, as well as numerous flight and engine instruments. The conditions necessitated constant engine changes; shortly after the group's arrival the only dozen spare engines at Nadi were soon depleted, and subsequent engine changes made at Plaine des Gaiacs soon exhausted New Caledonia's remaining stock.

On 2 October 1942 the group flew its most obscure SOPAC mission to Kapingamarangi Atoll (Greenwich) Island, which was used as an IJN flying boat staging base. This remote atoll was 850 miles northwest of Henderson Field and stretched the endurance of the Fortress, but the mission was nonetheless successful.

Throughout October and November 1942, the IJN operated destroyers on nocturnal supply runs to Guadalcanal which became known as the "Tokyo Express". During this critical period the 11th BG developed a *modus operandi* whereby a quartet departed daily at dawn from Espiritu Santo and searched out to a distance of 1,000 miles before landing at Henderson Field for an overnight stop. The next morning, they departed Henderson Field to cover a 450-mile arc up The Slot sometimes approaching close to Rabaul. This was efficient coverage but like so many other reconnaissance missions it was subject to the weather and visibility.

The 11th BG B-17 era ended when it returned to Hawaii in March/April 1943 to convert to the B-24 Liberator. The group logo was a shield with three grey geese, crested by a solitary flying goose, hence the nickname *The Grey Geese*.

19th Bombardment Group

The 19th BG's legacy stems back to 1927 when it was formed as the 19th Observation Group, prior to being redesignated as a bombardment group in 1929, then a heavy bombardment group in 1939. It was equipped with B-17s in 1941, moving to the Philippines in September of that year led by commander Major David Gibbs. However, Gibbs was soon replaced by Major Emmett O'Donnell on 12 December 1941. At the outbreak of war, the 19th BG comprised the 14th, 28th, 30th and 93rd BS, with the 435th BS later assigned in March 1942.[1]

On 8 December 1941 (7 December in the US) Japanese bombers and fighters attacked Clark Field, costing the 19th BG significant casualties and losses of B-17s on the ground. The 93rd BS, then on maneuvers at Del Monte, missed the attack. Supplies and headquarters were hastily moved from Clark Field to comparatively safe points nearby, and planes not badly damaged were repaired and dispatched to Del Monte.

---

[1] This particular 14th BS never operated in Australia as the unit remained extant in the Philippines, with USAAC ground personnel serving as makeshift infantry. Refer to chapters 3 and 9 in respect to subsequent confusion over the use of the associated 14th RS designation.

In subsequent weeks the 19th BG conducted numerous shuttle missions between Australia and Del Monte, using its bombers as transports between combat missions. They also evacuated troops and VIPs including General Douglas MacArthur and his family. During this busy time the bombers were often conducting two missions per day.

The first 19th BG B-17s to be based in Australia were fourteen B-17D/Es evacuated from Del Monte with the first arriving at Batchelor in the Northern Territory on 17 December 1941. On New Year's Day 1942, eleven B-17s flew to Malang in Java, to target Japanese shipping. These were reinforced by small numbers of B-17E reinforcements. All of the Fortresses were withdrawn to Australia at the end of February 1942 by which time Lieutenant Colonel Cecil Combs had become the new 19th BG commander. By this time most of the bombers were extremely war weary and were sent south to Laverton in Victoria for overhaul.

From March 1942 the 19th BG underwent a series of leadership changes. Lieutenant Colonel Kenneth Hobson was appointed commander on 14 March 1942 but only a month later he was replaced by Lieutenant Colonel James Connally who was in turn replaced by Lieutenant Colonel Richard Carmichael on 10 July.

During this period 19th BG Fortresses were flying missions against Japanese targets in New Guinea, using bases in Queensland and staging via Port Moresby. From 7 to 12 August 1942 the group struck Rabaul to distract the Japanese from attacking the US invasion of Guadalcanal. Captain Harl Pease Jr was posthumously awarded the Medal of Honor for his actions on the 7 August Rabaul mission when his Fortress fell behind and was shot down. The 19th BG returned to the US in late 1942 and was replaced in the New Guinea theatre by the 43rd BG.

43rd Bombardment Group

The 43rd BG was activated on 15 January 1941, and trained on a mixture of B-17, B-18, A-29 and LB-30 aircraft. The first echelon of support personnel arrived in Australia aboard the liner *Queen Mary* on 28 March 1942, however, the aircrew led by Colonel Roger Ramey did not arrive until much later to commence training with the departing 19th BG. The 43rd BG then took over much of that unit's Fortress inventory when the 19th BG returned to the US. The 43rd BG comprised the 63rd, 64th, 65th and 403rd BS.

The 43rd BG Fortresses hit New Guinea targets from Australian bases, staging through Port Moresby, until January 1943 after which it moved operations to Port Moresby. During its time in Australia, it operated from a variety of northern Australian bases including Daly Waters, Iron Range, Torrens Creek, Fenton and Mareeba. On 23 November 1942 a detachment deployed to Milne Bay in New Guinea, and from Port Moresby the group pioneered skip bombing. This technique was subsequently used in shipping strikes including during the Battle of the Bismarck Sea in March 1943.

The most senior Fifth Air Force officer ever lost in combat was aboard a 43rd BG B-17 on 5 January 1943. This was Brigadier General Kenneth Walker, Commander of Fifth Bomber Command, who was shot down during a daylight raid on Vunakanau airfield. Two crew were later captured, however, the fate of the rest of the crew and the resting place of the bomber remains unknown.

The 43rd BG Fortresses mainly flew missions against New Guinea targets, which included supporting ground forces, but with a key focus on Rabaul which was often attacked at night. On 30 June 1943 the group lost its first Fortress to a night fighter over Rabaul.

Lieutenant Colonel John Roberts replaced Ramey as 43rd BG commander on 30 March 1943, and then Colonel Harry Hawthorne took over from 24 May 1943 and saw out the group's B-17 era. The group later named themselves *Ken's Men* in honour of Lieutenant General George Kenney, the Fifth Air Force Commander. The 43rd BG commenced conversion to the B-24 Liberator towards the end of 1943 and had a motto of "Willing, Able, Ready" with the acronym of "WAR".

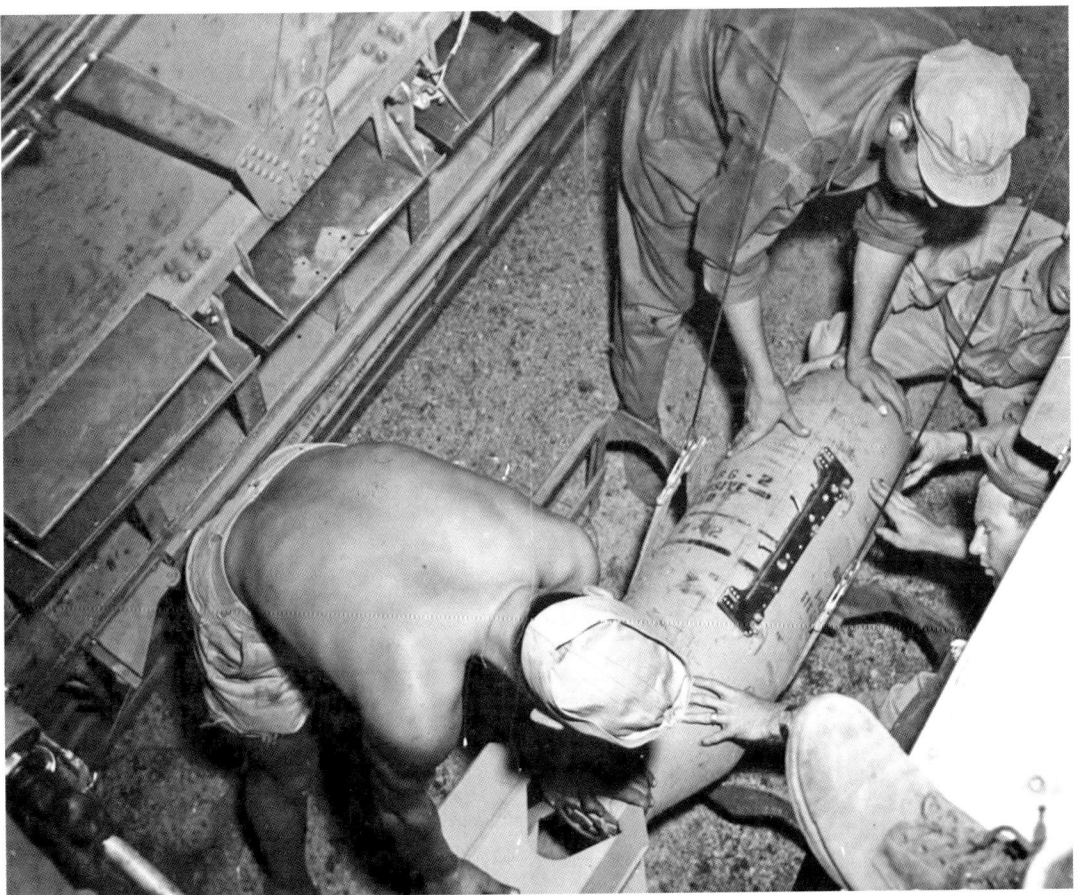

Ground personnel secure a bomb in the bomb bay of a 19th BG B-17E at Mareeba, Queensland, in mid-1942.

Taken just after its February 1942 arrival in Australia, 41-2416 has had its serial painted over in Hawaii and it has not yet been re-applied (see Profile 1).

B-17E 41-2429 landing at Nadi in February 1942 with the Batilamu Range in the background. This Fortress was later lost on 7 August 1942 over Rabaul when flown by Medal of Honor awardee Captain Harl Pease. This bomber also had its serial mistakenly painted over during application of the HAD scheme in Hawaii but lacks red and white rudder stripes.

CHAPTER 3
14th Reconnaissance Squadron

This unique *ad hoc* squadron conducted the first attack by USAAF bombers in New Guinea, albeit under USN command. In early 1942 Vice Admiral Wilson Brown was assigned the command of a dozen USAAC B-17Es to cover his Task Force 11 centred around his flagship, the carrier USS *Lexington*, as it headed for Rabaul. The bombers were commandeered from Hawaii's Seventh Air Force at short notice: six from the 88th RS, one from the 38th RS and five from the 23rd BS. The unit was known informally as the "Carmichael Detachment" after their commanding officer Major Richard Carmichael, a 28-year-old West Point graduate. After arrival in Australia the unit was designated as the 14th RS, but it was soon realised that the 14th BS was still extant in the Philippines serving as makeshift infantry. The unit was then briefly referred to as the 40th BS, before being formally redesignated as the 435th BS (see below). The squadron's fundamental objective was to protect Australia's shipping supply lines from the US.

Seven of the detachment's B-17Es were painted in the Hawaii Air Depot scheme: serials 41-2408, -2416 *San Antonio Rose*, -2421, -2429, -2430, -2432 and -2434. The other five were camouflaged in standard in Olive Drab: serials 41-2435, -2438, -2440, -2446 -2447. The distinctive red and white rudder stripes on the HAD bombers were painted over shortly after arrival in Australia.

After departing Hickam Field, in mid-February the Fortresses arrived at Nadi airfield on Fiji's main island of Viti Levu. From this RNZAF-managed facility, they conducted several extended search and reconnaissance missions, some up to 700-miles range. The Fortresses then flew to RAAF Garbutt, Townsville, over 18 to 20 February 1942, transiting via Plaine Des Gaiacs in New Caledonia.

Only a few days later Carmichael was ordered to bomb Rabaul, the first such attack by USAAC bombers in New Guinea. The plan was to strike shipping in Simpson Harbour at dawn, involving all available bombers, but the mission was beset with difficulties and only eight were ready to depart in the early hours of 23 February. One of these suffered a ground collision and was later scrapped (*San Antonio Rose*) and another soon returned to base with engine trouble, leaving just six heading for the target. Another turned back in the face of severe weather, while over the target the five remaining bombers were intercepted by Japanese fighters. As a result of damage incurred and fast running out of fuel, First Lieutenant Fred Eaton's bomber crash landed in a swamp in New Guinea while trying to reach Port Moresby.

The Fortresses continued to operate as the Carmichael Detachment until 14 March 1942 when it was reassigned to the command of the 19th BG briefly as the 40th RS, before around a month later it was redesignated as the 435th BS. The 435th BS is detailed in Chapter 9.

When they left Hawaii, the only 14th RS bomber named was *San Antonio Rose*. Other publications claim that several others had nose art, but they have wrongly assumed that names later applied during the latter 19th and 43rd BG eras were extant when they left Hawaii. The 14th RS had no logo.

Profile 1   B-17E serial 41-2416 *San Antonio Rose*

First Lieutenant Frank Bostrom ferried this bomber to Australia from Hawaii. The bomber was written off following a ground collision while preparing for the 23 February Rabaul mission. The name was accompanied by the artwork of a dancing Mexican girl. This HAD bomber exceptionally had its original serial number mistakenly painted over in Hawaii, so the last four digits "2416" were later reapplied In Australia, as profiled.

Profile 2   B-17E serial 41-2446

Colloquially know today as *Swamp Ghost*, this bomber in fact was unnamed at the time of its loss on 23 February 1942 during the first USAAC raid in New Guinea. On this mission First Lieutenant Fred Eaton's bombs would not release over Rabaul, so he made a second pass when he was attacked by Claudes and Zeros. Eaton was pursued from the area and experienced fuel loss because of combat damage. He force-landed in the Agaiambo Swamp and all crew survived, taking about a week to find their way out of the swamp. The bomber was salvaged in 2006 and is currently on display at the Pearl Harbor Aviation Museum.

Profile 3   B-17E serial 41-2432

First assigned to the 23rd BS, on 7 December 1941 this bomber experienced the Pearl Harbor attack. On 16 January 1942 it was one of three B-17Es assigned to assess sea routes from advance bases and ventured as far as Fiji before returning to Hawaii, where it received the HAD scheme. In February it arrived in Australia as part of the Carmichael Detachment. It later served with the 63rd BS named *The Last Straw* before being converted into an armed transport (see Profile 105).

Profile 4   B-17E serial 41-2435

This Fortress originally served with the 30th BS pre-war in the US and was destined for the Philippines before it was held in Hawaii. It was reassigned to the Carmichael Detachment and arrived at RAAF Garbutt on 19 February 1942. Two months later it was reassigned to the 28th BS and was later lost with that unit.

Profile 5   B-17E serial 41-2430

This B-17E was damaged by DC-3 airliner VH-ACB which taxied into it on evening of 18 February1942 after it arrived at Archerfield, Queensland, from New Caledonia. The Fortress sustained damage to its starboard wing, tail and rear fuselage, and subsequent repairs delayed its return to the 14th RS in Townsville until 27 February 1942. Hence it missed out on the inaugural Rabaul raid, and later became *Naughty But Nice* when assigned to the 65th BS (see Profile 59).

B-17E 41-2520 Jap Happy, as illustrated in Profile 7, at Guadalcanal decorated with the pre-war 23rd BS emblem.

B-17E 41-9222 L'il Nell, the subject of Profile 9, cruises up The Slot towards Buin.

CHAPTER 4
23rd Bombardment Squadron

Assigned to the 5th BG, when war was declared the 23rd BS was based in Hawaii with the B-17E Flying Fortress. Throughout most of 1942 the squadron flew patrol missions around Hawaii before the end of the year when it moved to Espiritu Santo to support the Solomons campaign. From 31 March to 24 August 1943, it deployed to Guadalcanal from where it conducted numerous missions over southern Bougainville targets including at night. During these missions it lost only two B-17Es, both to J1N1 Irving night-fighters: *Tokyo Taxi* on 19 July 1943 and then *De-icer* a week later. Meanwhile when the 394th BS was disbanded all of its B-17Es were transferred to the 23rd BS, making it the final B-17 bombardment squadron in the SOPAC theatre. However, this period was limited. On 24 August 1943 the squadron's B-17s returned to Espiritu Santo for rest and recuperation in preparation for transition to the B-24 Liberator.

During the early B-17 era the 23rd BS commander was Major Laverne "Blondie" Saunders, who had assumed the position in December 1938. When Saunders was promoted to command of the 11th BG in July 1942, he was replaced by Major George Blakey. The next squadron commander was Major Russell Dellinger who saw out the unit's B-17 era. The 23rd BS lost only two Fortresses during its SOPAC deployment, both to night fighters, as noted above.

The 23rd BS used a pre-war logo of a skeleton thumbing its nose and holding a bomb, as shown at the top of page 30. Several B-17s (including Profile 7) applied the logo on the nose of their bombers to complement their nose art.

Crewmen pose in front of B-17E 41-9222 *L'il Nell*, as depicted in Profile 9. Note the four antenna protrusions of the SCR-521 long-wave ASV search radar installation.

Profile 6   B-17E serial 41-9099 *Miss Betty*

This Fortress survived its combat tour with the 23rd BS and returned to the US on 20 March 1945. In August 1945 it was sent to Stillwater, Oklahoma, for storage, before it was scrapped the following year.

Profile 7   B-17E serial 41-2520 *Jap Happy*

This B-17E was transferred to the 23rd BS via the Hawaii Air Depot on 28 May 1942. On the night of 26-27 July 1943 just after a pass over Kahili, it was isolated by searchlights and shot up by a J1N1 Irving night fighter. It returned to base and was then named *Jap Happy* to celebrate the fact it had safely returned its crew to Henderson Field. It served as a transport from late 1943 onwards and eventually returned to the US, leaving Espirito Santo on 2 September 1944.

Profile 8   B-17E serial 41-9153 *Tokyo Taxi*

Just before midnight on 18 July 1943 this was one of nine B-17Es which departed Guadalcanal to attack Buin. Flying in the lead position over the target at 14,000 feet, the bomber was cornered by searchlights and shot down by a J1N1 Irving night fighter in early hours of the next morning. It was last seen aflame eight miles north of the target and remains missing. The No. 251 *Kokutai* Irving was shot down by the bomber's return fire. The name *Tokyo Taxi* was applied to both sides of the nose.

Profile 9   B-17E serial 41-9222 *L'il Nell / Knucklehead*

This bomber arrived at Espiritu Santo in July 1942 while serving with the 431st BS and was named *Knucklehead*. There it was one of several B-17Es fitted with SCR-521 long-wave ASV search radar, installed under the port wing with a nose antenna and extra dorsal aerial. In February 1943 it was assigned to the 23rd BS and Captain Anthony Lucas who named the bomber *L'il Nell* after his wife Nell. Lucas also painted the Disney cartoon character *Thumper* just above the name. Following over a year of combat service in the Solomons the bomber was ferried back to the US in October 1943. There it was refitted as a reconnaissance RB-17E and operated from Wright Field commencing 2 June 1944.

Profile 10   B-17E serial 41-2467 *San Antonio Rose*

This B-17E was received by the 23rd BS at Hickam on 16 February 1942 and was among those camouflaged in the HAD scheme. It arrived at Espiritu Santo on 25 October 1942 where it was named *San Antonio Rose* and decorated with artwork of a calendar girl wearing a top hat. Following its 23rd BS combat service, *San Antonio Rose* was ferried back to the US in October 1943.

B-17F 41-24457 *Aztec's Curse*, as illustrated in Profile 11, at Henderson Field.

The ad hoc nature of the HAD scheme is exemplified here on this 23rd BS Fortress where considerable segments of overspray are evidenced on the wing undersurfaces.

CHAPTER 5
26th Bombardment Squadron

The 26th BS was among the 11th BG squadrons that deployed to the South Pacific in July 1942. Under the command of Major Allan Sewart, the squadron was initially based at Efate. In Hawaii the squadron's Fortresses had been modified with the installation of additional fuel tanks in the forward fuselage, just behind the radio compartment. Filling the tanks to capacity limited the bomber to carrying 2,000 pounds of bombs, usually in either 4 x 500-pound or 20 x 100-pound configurations. One engineer per bomber was all that could accompany the bombers to Efate, meaning that a disproportionate share of service and maintenance duties in the early days fell to the aircrews themselves.

The 26th BS's involvement in the early Guadalcanal campaign is landmark. On 30 July 1942 Sewart flew the first Fortress to land at the newly constructed Palikulo airfield on Espiritu Santo, and the 26th BS shouldered a heavy burden of early missions as the other three 11th BG squadrons were home based in New Caledonia and Fiji. Initially photo reconnaissance missions were flown over Solomons on 23 and 25 July, and on both occasions A6M2-N Rufes attempted interception.

On 31 July the 11th BG made its first bombing raid against Guadalcanal, with the eight-bomber formation including six 26th BS B-17Es. On 3 August a pair of 26th BS Fortresses bombed Tulagi from 6,000 feet, then the following day a trio of 26th BS bombers was involved in further attacks. These were intercepted by six Rufes and B-17E 41-9218 was lost in a mid-air collision with one of the floatplane fighters.

Subsequent to the US invasion of Guadalcanal, during the Battle of the Eastern Solomons on 24 August 1942 a formation of 11th BG B-17Es departed Palikulo to attack the carrier *Ryujo*, then about 650 miles distant. The attack was authorised knowing all participants risked a dangerous return night landing at Palikulo's primitive airfield. The first flight of three bombers included one from the 26th BS, while the second flight of four was led by Major Allan Sewart and included three 26th BS machines. The first flight attacked from 12,000 feet in the late afternoon, although it is unclear if any of the bombs hit the carrier, while the second flight struggled to find targets due to poor light.

On the return flight the Fortresses battled a torrential downpour as they approached Espiritu Santo in darkness. First Lieutenant Robert Guenther flying B-17E 41-2610 hit palm trees when an engine failed as he was attempting to land, with the crash killing Guenther and four others. Meanwhile Sewart wisely diverted the second flight around the weather to Efate, right at the limit of their fuel endurance. After landing safely at Efate all four returnees were found to have sustained damage from anti-aircraft gunfire.

An unusual incident occurred in the early morning of 23 October 1942 when submarine *I-7* surfaced off Espiritu Santo to shell Palikulo airfield. It only fired six rounds before diving in response to return fire from a shore battery. A 26th BS B-17E, 41-9076, tried unsuccessfully to find the submarine, however, on return to Palikulo its brakes failed and it collided with a parked fighter. The bomber was later repaired and returned to service.

On 18 November 1942 a formation of ten B-17s set out from Henderson Field to bomb shipping in the Shortlands, led by 26th BS commander Major Sewart in B-17F serial 41-24531. The bombers were intercepted by a mix of Zeros, Rufes and Petes, with Sewart's lead aircraft targeted in head-on attacks. After Sewart was killed and his co-pilot mortally wounded, the 11th BG commander, Colonel Laverne "Blondie" Saunders, who was riding aboard as an observer, took the controls. With his two left engines shot out, Saunders nursed the crippled bomber away from the area but was soon forced to ditch off Vella Lavella. The surviving crew took to a dinghy and were assisted by a coastwatcher and friendly natives. They were rescued by a PBY the following day (Sewart's B-17F 41-24531 is depicted in Profile 15).

Throughout the rest of the year 26th BS Fortresses continued to hit numerous targets in the Solomons, and in January 1943 the unit sent a detachment to Port Moresby's Seven-Mile 'drome for a week to assist Fifth Air Force efforts in New Guinea. At the end of February 1943, the squadron's Fortresses were redistributed among 5th BG units, as the squadron prepared to transition to the B-24D Liberator. By this time the 26th BS had been reduced to just five airworthy B-17Es. It lost six Fortresses in the South Pacific to combat and accidents.

The 26th BS logo was a blue and orange clenched fist in front of a shield featuring the same colours but on opposite sides. It is shown on the top of page 36.

B-17E 41-9217 The Globe Trotter, the subject of Profile 12, seen at Plaine des Gaiacs following its 27 October 1943 accident.

Retired SOPAC bombers never died, at least not until they were scrapped. This is B-17E 41-2524 (as shown in Profile 13) in early 1944 when serving with the 2135th Base Gunnery School at Tyndall Field, Florida. The fin's leading edge retains the white paint previously applied by the 26th BS ground crew at Espiritu Santo.

A group of 26th BS B-17Es at Tontouta in February 1943 awaiting refurbishment before being redistributed between other 5th BG squadrons.

26th Bombardment Squadron

Profile 11 B-17F serial 41-24457, *Aztec's Curse*

This "F" model was among the first delivered to the South Pacific, arriving at Palikulo on 21 August 1942 where it was named *Aztec's Curse*. In February 1943 the bomber was transferred to the 31st BS with which it was badly damaged when it ground looped after brake failure on landing on 23 April 1943. A ground tug subsequently further damaged the rear fuselage when trying to tow the bomber.

Profile 12 B-17E serial 41-9217, *The Globe Trotter*

This bomber was serving with the 431st BS when it arrived at Nadi, Fiji, on 22 July 1942. On 17 August 1942 it landed on the recently captured Henderson Field, becoming the first USAAF bomber to land there. It was subsequently transferred to the 26th BS for use as a transport when the 11th BG commenced transition to the B-24D Liberator, however, it was badly damaged when it had a landing accident at Plaine des Gaiacs, New Caledonia, on 27 October 1943. Note at this stage the leading edge of the fin had been painted white (not left as natural metal finish) to cover the area where the anti-ice boot had been removed. The accident damaged the nose and both left engines, however, the B-17E was repaired and returned to service as a transport.

Profile 13 B-17E serial 41-2524

This 26th BS B-17E arrived at Efate on 22 July 1942. On 2 August 1942 during a mission to Tulagi it was damaged by Rufe floatplanes. On a subsequent mission against shipping at Tonolei Harbor on 18 November 1942 it was intercepted by Zeros which made numerous head-on passes, and despite knocking out two engines the bomber was able to get home. It was ferried back to the US in December 1943.

Profile 14 B-17E serial 41-2433, *Miss Fit*

When serving with the 23rd BS, this Fortress was grounded in Fiji for several weeks awaiting parts, so it earned the name *Miss Fit*. After returning to service in July 1942 it was reassigned to the 26th BS and shortly after arriving at Palikulo it was fitted with an SCR-521 ASV search radar. During a patrol on 21 November 1942, it skirmished with a Mavis flying boat, and was forced to return to Henderson Field after one engine was knocked out. The bomber was transferred to a service squadron in August 1943, prior to returning to the US in early 1944.

Profile 15 B-17F serial 41-24531

This was the bomber in which 26th BS commander Major Allan Sewart was killed on 18 November 1942, as described on page 34. After combat with Japanese fighters over the Shortlands, it was badly damaged prior to being ditched by the 11th BG commander Colonel Laverne Saunders who was riding aboard as an observer.

B-17E 41-2638 "Chico", the subject of Profile 17, at Seven-Mile 'drome, Port Moresby.

B-17E 41-2472, as illustrated in Profile 16, in a revetment at Seven-Mile 'drome, Port Moresby, when it was serving with the 43rd BG as a transport. Its new name Guinea Pig is on the nose.

CHAPTER 6
28th Bombardment Squadron

Part of the 19th BG, the 28th BS was based in the Philippines when war broke out. Following the initial series of attacks against airfields on Luzon in early December 1941 the squadron's Fortresses were dispatched to Del Monte Field on Mindanao. Meanwhile, most of the ground echelon remained in the Philippines and fought as infantry to defend the Bataan Peninsula, costing many their lives. From Del Monte a handful of Fortresses began reconnaissance and limited attacks against Japanese ships and troops until 24 December 1941. Lacking spares and maintenance support, the Fortresses moved south to Batchelor near Darwin.

In January 1942 the 28th BS staged Fortresses to Java in order to support Allied efforts during the NEI campaign. The Dutch capitulation several weeks later forced the squadron to again withdraw to Australia in March. It participated in the Battle of the Coral Sea in May 1942, and often bombed enemy shipping in New Guinea and Rabaul, staging through Port Moresby.

The unit was reformed as part of the Fifth Air Force in August 1942, after receiving replacement B-17Es and crews from the US.

The 28th BS lost two B-17Es to combat in New Guinea, and in late 1942 Washington decided that the entire 19th BG would be withdrawn and replaced by the 43rd BG. Therefore, in early December 1942 the remaining 28th BS inventory of B-17Es commenced transfer to the 43rd BG, after which the squadron's long-serving personnel returned to the US. On 30 December 1942 the unit was reclassified at Pocatello, Idaho, as a replacement training unit and re-equipped with B-17Es. Major Elbert "Butch" Helton commanded the squadron from 14 March 1942 until its return to the US.

The 28th BS logo emulated an original 1924 design, featuring the head of a Mohawk Indian underneath a blue diamond. It is shown at the top of page 40.

The wreck of B-17E 41-9234, as depicted in Profile 20, seen in 1953 ten years after it crash-landed. Visible are the RAF fin flash and other markings including the red centre roundel.

Profile 16   B-17E serial 41-2472 *Ole Miss II / Guinea Pig*

This bomber arrived in Java via the African delivery route on 15 January 1942, having been named *Ole Miss II* by Captain Fred Key prior to leaving the US. The bomber's original Sperry remote turret was replaced with a Bendix turret around mid-1942, and a few months later it was reassigned to the 65th BS. Boeing claimed in a 1944 newsletter that this Fortress flew the most missions of any B-17 in the theatre, and in Hawaii in late 1943 it had been decorated with 208 mission markers, 23 Zero kills and three ships sunk. The artwork, however, was creative fiction, similar to many such scoreboards applied at short notice on long-serving Fortresses returning stateside. In fact, the 43rd BG had used the war-weary aircraft mainly for transport purposes and had renamed it *Guinea Pig*, a disparaging name, reflecting the frequent maintenance problems it presented.

Profile 17   B-17E serial 41-2638 *"Chico" / I'm Willing*

This Fortress arrived in Australia on 30 April 1942 and shortly afterwards it was named *"Chico"* at Mareeba. It was transferred to the 65th BS on 20 January 1943, and its new ground crew painted a Vargas calendar girl under the name *Chico*. Some months later it was reassigned to the 63rd BS where it was renamed *I'm Willing* in Gothic letters as illustrated. The bomber was ferried back to the US in November 1943.

Profile 18   B-17E serial 41-2637 *"Gone Forever"*

This bomber was delivered to the 28th BS on 18 April 1942. It was named *Gone Forever* by the ground crew, however, it is unclear when the reclining nude was added, possibly by the 64th BS to which it was transferred in December 1942. On the left side of the nose another similar reclining nude was painted, and above which the name *Ready Betty* was applied in small letters. The bomber flew its final mission in October 1943 before it was flown back to the US.

Profile 19   B-17F serial 41-24428 *Miss Carriage*

This Fortress was among the first batch of new "F" models delivered to the 28th BS in late August 1942. The bomber had barely entered service when on 5 September 1942 it was returning to Mareeba at night from New Guinea, and the right landing gear collapsed on landing. The brand-new aircraft was subsequently written off.

Profile 20   B-17E serial 41-9234, RAF serial FL461

This Fortress was contracted to RAF Coastal Command but was hastily reassigned to the USAAF before arriving in Australia in August 1942 sporting British camouflage. Following assignment to the 28th BS its nose compartment was modified to house B-17F-style gun windows, and it flew several missions with the squadron before it was handed over to the 63rd BS in November 1942. On 8 January 1943 it was damaged by anti-aircraft fire while bombing a Japanese convoy. Unable to climb over mountains, it was forced landed near Kaisenik village. Today it is the most famous existing wartime wreck in Papua New Guinea.

B-17E 41-2640 *Tojo's Physic*, as illustrated in Profile 24, with its crew at Mareeba. The bomber was written off following a ground collision at Horn Island on 27 July 1942.

B-17F 41-24403 *Blitz Buggy*, the subject of Profile 25, at Mareeba under a partly constructed hangar.

CHAPTER 7
30th Bombardment Squadron

Part of the 19th BG, the 30th BS experienced the early setbacks of the Philippines campaign before moving its surviving B-17s to Java in early 1942. Meanwhile, much of the ground echelon remained on Luzon and fought as infantry to defend the Bataan Peninsula, costing many lives.

On 5 March 1942 the 30th BS's air echelon left Java for Batchelor in Australia's Northern Territory. From here supply and evacuation flights were conducted to the Philippines. Meanwhile those B-17Es which required much-needed maintenance were sent down to RAAF Laverton, Melbourne, where RAAF engineers assisted repairs.

After a three-week stay at Batchelor the 30th BS then moved to Cloncurry in Queensland. Its first combat loss in New Guinea took place on 25 April 1942 when B-17E 41-2505 disappeared after departing Port Moresby in the early morning for a bombing mission against Rabaul. It crashed into Mt Obree at 0446 and was not located until post-war. On 13 May 1942 the squadron moved to Longreach where it remained for two months before moving further north to Mareeba.

Following the late 1942 decision to withdraw the 19th BG from the theatre, in November the remaining 30th BS inventory of B-17Es was transferred to the 43rd BG, after which the squadron's long-serving personnel returned to the US. Major Thomas Steed as squadron commander oversaw the early operations including the evacuation from the Philippines and Java. He was replaced by Lieutenant Colonel Jack Wood on 21 August 1942 who saw out the remainder of the squadron's time in Australia and New Guinea. The 30th BS lost fourteen Fortresses to combat and accidents in the SWPA theatre.

The 30th BS logo was a walking cartoon policeman carrying a truncheon, as shown at the top of page 44. This design was never officially approved, however, and was replaced in 1954 with a different design.

B-17E 41-2434 seen in HAD camouflage during a 1942 visit to Ohakea airfield, New Zealand. This was one of the original 14th RS bombers that arrived in Australia in February 1942 (see Chapter 3). It was subsequently transferred to the 30th BS but crashed during a test flight from Townsville on 16 August 1942.

PACIFIC PROFILES

30th Bombardment Squadron

Profile 21   B-17E 41-2659 *Frank Buck*

Ferried to Australia via the South Pacific route, this bomber was assigned to the 30th BS on 13 July 1942. It was named *Frank Buck* in honour of the 1930s radio serial about a fictitious adventurer, *Frank Buck Bring 'em Back Alive*. During a mission against Vunakanau on the night of 15-16 September 1942 the bomber was damaged by anti-aircraft fire and faced bad weather on the return flight forcing it to land on a beach southeast of Port Moresby. It was successfully flown off the next day. After the 30th BS returned to the US, the bomber served as a transport before it was scrapped in July 1943.

Profile 22   B-17E 41-9235 *Clown House*

Originally built for an RAF order, this bomber was instead diverted to the USAAF and arrived in Mareeba in August 1942 where it was operated by the 30th BS. During a night mission from Mareeba, staged via Port Moresby against shipping in Tonolei Harbor on 28 October 1942, the crew jettisoned their bombs when they could not locate the target. On the return flight and low on fuel the Fortress was ditched near Cooktown. The crew languished at sea in a dinghy for two days until rescued by Aboriginals.

Profile 23   B-17E 41-2635

Thei B-17E was delivered to the 30th BS in mid-1942. On the night of 1 November 1942 with First Lieutenant John Hancock at the controls, the Fortress failed to return from a night raid against shipping in Tonolei Harbor, Bougainville. The weather on the journey was marked by low cloud and rain, and in poor visibility the bomber had crashed into a ridge near Milne Bay, killing all aboard.

Profile 24   B-17E 41-2640 *Tojo's Physic*

This bomber arrived in Australia in April 1942 where it was assigned to the 30th BS. At Horn Island airfield on 27 July 1942 the pilot was preparing for take-off when the bomber's nose was torn off by the wing of B-17E 41-2460, which had just landed after bombing Lae. Poorly lit runways at Horn Island in the dawn light contributed to the accident. Fortunately, nobody was killed in the accident but both bombers were subsequently written off.

Profile 25   B-17F 41-24403 *Blitz Buggy / The Old Man*

This "F" model was delivered to the 30th BS at Mareeba on 26 June 1942 where it was named *Blitz Buggy*, and it is profiled as it appeared in service with the squadron two months later. When the 30th BS returned to the US, it was reassigned to the 65th BS. In early 1943 the original name was painted over and replaced with artwork depicting a cheerful Uncle Sam and the new name, *The Old Man*, as illustrated. After the bomber was retired from combat in August 1943 it served as the personal transport of Lieutenant General Ennis Whitehead. *The Old Man* was finally scrapped at Clark Field in July 1948.

B-17E 41-9124 Buzz King, as depicted in Profile 27, at Tontouta, New Caledonia.

The first detachment of 31ˢᵗ BS B-17Es arrived at Palikulo, Espiritu Santo, on 30 November 1942. This is a Fortress from the detachment departing the airfield the following month.

CHAPTER 8
31st Bombardment Squadron

Part of the 5th BG, in late 1941 the 31st BS was part of the Seventh Air Force based at Hickam Field operating both the B-18 Bolo and B-17D Fortress. The squadron had a rich pedigree, deriving from the 31st Aero Squadron which had been formed during the Great War in 1917, operating Nieuport 21s from Etampes in France. It had later been reconstituted and redesignated the 31st Bombardment Squadron in May 1923 where it operated from Hamilton Field in California. It took up station at Hickam Field, Hawaii, in February 1938 where it first operated B-18 Bolos before relocating in May 1942 to Kipapa Field on Oahu. The squadron participated in the June 1942 Midway campaign during which it lost B-17E 41-9212, prior to its first B-17E detachment arriving at Palikulo, Espiritu Santo, on 30 November 1942.

On 17 January 1943 it moved operations to Henderson Field, Guadalcanal, before moving again to nearby Carney Field. On 13 January 1943, the entire parent 5th BG was transferred from the Seventh Air Force and reassigned to the newly created Thirteenth Air Force. The latter was activated on this date as an initiative to coordinate the handling of all USAAF units in the SOPAC theatre, all of which now fell under the command of Brigadier General Nathan Twining.

Shortly after its move to Henderson Field on 17 January 1943 the 31st BS began converting to the B-24D Liberator, redeploying operations to Munda in February 1944. Nonetheless, a handful of its original B-17Es were still operating as late as June 1943 during transition to the B-24D Liberator. The squadron commander who led the unit to the South Pacific was Major George Glober. He was replaced by Major FT Brady on 30 March 1943.

The 31st BS lost two Fortresses to combat in the SOPAC theatre. One of these remains MIA over Buin, and B-17E 41-9124 *Buzz King* (see Profile 27) was destroyed during a bombing raid when parked at Henderson Field on 23 March 1943. The nature of operations at the time meant the squadron's bombers were often left fully loaded with bombs and fuel ready for night mission departures.

The 31st BS logo was approved in 1934 and featured a skull and crossbones on a black triangle. It is illustrated above the profiles on page 48.

Profile 26   B-17E 41-2525 *Madame X*

After participating in the Battle of Midway, this bomber arrived at Espiritu Santo in November 1942. On 10 June 1943 it departed Carney Field on Guadalcanal for an early morning mission to bomb Buin. Arriving over the target at dusk, pilot First Lieutenant Richard Snoddy reported he had a faulty artificial horizon and wanted to abort. He was instructed to continue using needle, ball and airspeed but replied his skill at doing such was limited. *Madame X* failed to return and is believed to have gone into a spiral dive in the vicinity of the target area.

Profile 27   B-17E 41-9124 *Buzz King*

Shortly after arriving in the South Pacific in November 1942 this Fortress was field-modified at the Thirteenth Air Force Depot at Tontouta, New Caledonia, with twin 0.50-inch calibre machine guns mounted just behind the radio compartment. It was destroyed by bombs dropped by No. 705 *Kokutai* G4M1 Bettys when parked at Henderson Field on 23 March 1943. The bomber was loaded with eight 500-pound bombs and was full of fuel ready for a night mission to bomb Buin.

B-17E 41-2525 Madame X, the subject of Profile 26, undergoing repairs at Espiritu Santo. Note the Midway battle star alongside the mission markers.

B-17E 41-2656 *Chief Seattle*, as depicted in Profile 32, at Seven-Mile, Port Moresby, in the first week of August 1942 shortly after its trans-Pacific ferry flight.

B-17E 41-2408, the subject of Profile 28, shows off its HAD camouflage at Cairns in late 1942.

CHAPTER 9
40th Reconnaissance Squadron and 435th Bombardment Squadron

Refer to Chapter 3 for the origins of the 14th Reconnaissance Squadron, which arrived in Australia from Hawaii in February 1942. The unit was theoretically redesignated as the 40th RS on 14 March 1942 by the War Department, but this designation was never officially activated by the USAAC before the unit was instead redesignated as the 435th BS on 22 April 1942. At this time the unit joined the 19th BG.

Nonetheless, several Fortresses operated for just over a month under the banner and administration of the 40th RS, a situation that has often caused confusion. Subsequently, the 435th BS became known as the "Kangaroo Squadron" because it was the first USAAC squadron to properly operate in Australia, noting that parts of the original four 19th BG squadrons had temporarily been in the country from December 1941 before moving to Java.

On 26 March 1942 many experienced ground personnel recently evacuated from Java joined the squadron, bringing relief to the existing over-worked crews. The unit soon boasted a diverse cadre of some 400 men, representing all but three of the 48 states, and including several who were born overseas. When the original squadron commander, Colonel Richard Carmichael, was promoted to command the 19th BG, Major William Lewis assumed command.

The 435th BS played a key reconnaissance role in the Battle of the Coral Sea in May, however, arguably its most critical accomplishment was reconnaissance of early Japanese positions at Guadalcanal and Tulagi. On 18 June 1942 a 435th BS Fortress flew the first reconnaissance of Tulagi, then in early July the squadron commenced regular daily reconnaissance flights over the area from both Townsville and Port Moresby. The resulting photos facilitated planning for the subsequent Marine landings in August 1942.

The 435th BS flew its last mission in New Guinea on 10 October 1942. Thereafter its crew returned to the US with its surviving Fortress inventory reassigned among the 43rd BG. The unit's armament personnel were the first in the SWPA to design and install twin-0.50-inch calibre machine guns in the Perspex nose. The same engineers also mounted movie cameras in the waist and tail gun positions to acquire film for gunnery training. Upon its return to the US, the squadron was directed to compile a technical directive entailing a summary of lessons learned from its combat experience in New Guinea. While in the theatre, the 435th BS lost a total of eight Fortresses to combat and accidents.

The logo for the 40th RS was never approved and consisted of a fighting cock with boxing gloves against a blue background. The logo for the 435th BS was a leaping kangaroo in front of a large cloud, peering through a telescope grasped in the forelegs and holding a golden bomb with its tail. Both logos are shown at the top of page 52.

Profile 28   B-17E 41-2408

This Fortress was one of the original 14th RS bombers which arrived in Townsville from Hawaii on 19 February 1942. A month later it was one of the three B-17Es to participate in the evacuation of General Douglas MacArthur from the Philippines. When the 435th BS returned to the US, the bomber was reassigned to the 64th BS at Mareeba in November 1942. In late 1943 it was converted into an armed transport and served with the 46th TCS. The bomber was scrapped at Brisbane in 1945.

Profile 29   B-17E 41-2429 *"Why Don't We Do This More Often?"*

This Fortress flew into Hawaii during the Pearl Harbor attack and was subsequently painted in the HAD camouflage scheme before flying to Australia for service with the 435th BS. Captain Harl Pease and his crew flew this bomber in a 7 August 1942 raid over Rabaul, after their original machine suffered a faulty engine. Flying at the edge of the formation, the bomber was shot down by a large formation of Zeros. A photo of this bomber appears at the bottom of page 24.

Profile 30   B-17E 41-2447 *San Antonio Rose II*

Like 41-2408 (Profile 28), this Fortress was another of the original 14th RS bombers that arrived in Australia in February 1942. It was assigned to the "Royce Mission" to attack targets in the Philippines, and on 11 April 1942 it flew to Del Monte airfield on Mindanao. However, during the flight one engine became inoperative, and the bomber was grounded for the next morning's mission. Parked in the open, and thus exposed to aerial attack, it was bombed by a Pete floatplane.

Profile 31   B-17E 41-2652

This Fortress was delivered to the 40th RS on 22 April 1942 after being ferried across the Pacific. During the Battle of the Coral Sea on 7 May 1942 it became lost in darkness when returning from attacking Japanese warships. When it ran low on fuel the crew baled out, however, the bomber flew another 40 forty miles before crashing in northern Queensland. The crew landed unhurt, and all were rescued.

Profile 32   B-17E 41-2656 *Chief Seattle*

The bomber was named *Chief Seattle* after it was purchased by Seattle's citizenry as part of a war bond drive. It arrived in Australia on 2 August 1942, but barely a fortnight later on 14 August the bomber went missing during a reconnaissance of Gasmata and Rabaul. Japanese records show it was shot down by a *chutai* of nine Tainan *Kokutai* Zeros.

Profile 33   B-17E 41-2634 *Tex*

This Fortress arrived in Australia in early July 1942 where it was assigned to the 435th BS and named *Tex*. Shortly thereafter it had a SCR-521 surface search radar installed to detect enemy ships at night. In late 1942 it was reassigned to the 65th BS at Mareeba, where it was renamed *Red Moose Express* (see Profile 58).

Profile 34   B-17E 41-9207 *Flagship Texas No. VI / Strip-Straffer*

This Fortress was delivered to the 435th BS at Townsville on 8 July 1942. Just prior to its departure from the US it was equipped with an SCR-521 ASV search radar. However, this proved ineffective, and its drag reduced range, so it was removed in April 1943. The bomber was named by its original crew in honour of the first American Airlines DC-3 named *Flagship Texas*. The name *Strip-Straffer* on the other side had the boy wearing an American Airlines pilot's cap. It appears both artworks were applied by someone with a previous association with American Airlines, reflecting an in-house joke. After the 435th BS returned to the US in October 1942 the bomber was transferred to the 64th BS with which it was shot down over Rabaul on 1 June 1943.

An unidentified Olive Drab 435th BS B-17E looking towards the north side of Seven-Mile, Port Moresby, in 1942.

A 435th BS B-17E is refuelled at Cloncurry, Queensland, in May 1942.

B-17E 41-2444 was another HAD Fortress in service with the 42nd BS, seen here being serviced on Espiritu Santo.

CHAPTER 10
42ⁿᵈ Bombardment Squadron

A component of the 11th BG, on 18 July 1942 the first detachment of 42nd BS B-17Es led by Major Ernest Manierre departed Hawaii and staged to Plaine des Gaiacs, New Caledonia, via Christmas Island, Canton and Nadi. All nine Fortresses arriving on 23 July 1942 were: 41-9071 *Stingaree*, 41-9222, 41-2420 *Bessie Jap Basher*, 41-9216 *Alley Oop*, 41-2445 *So Solly Please*, 41-9151, 41-9155, 41-9213 and 41-2442 *Yokohama Express*.

Plaine des Gaiacs then boasted New Caledonia's longest runway at 7,000 feet, however, with no spare revetments large enough to accommodate the B-17s all were briefly detached to Koumac. The 42nd BS flew search missions from Koumac until the invasion of Guadalcanal, with most of the bombers having been fitted with additional fuel tanks at the Hawaii Air Depot. The Fortresses later returned to Plaine des Gaiacs commencing patrols as far as Indispensable Reef, just to the south of Guadalcanal.

So acute was the need for reconnaissance cover in the area that some missions borrowed cameras from VMO-251, an outfit fitted with reconnaissance Wildcats. On 26 July *So Solly Please* encountered two Jake floatplanes over Indispensable Reef, however, the parties were too distant to engage. Then on 31 July an advance detachment was sent to Efate so they could be close enough to photograph Guadalcanal itself. A month later, on 30 August, the 42nd BS received its first B-17F (see profile 36).

On 2 September 1942 a 42nd BS B-17E flown by Captain Richard Eberenz located a "tanker" off Santa Isabel. This was instead IJN patrol boat *PB-35*, sunk by a direct hit, which had embarked support crews from the seaplane tender *Sanyo Maru* to resupply the advanced base at Rekata Bay. Six days later Captain Robert Richards and his crew flying *Stingaree* were shot down by a Toko *Kokutai* H6K4 Mavis over Rendova Island.

On 2 November 42nd BS crews took their first leave in New Zealand, transported there by USMC R4Ds. Throughout January 1943 the squadron's Fortresses dropped supplies to US troops fighting the Japanese in Guadalcanal's foothills. Major Earl Hall replaced Major Ernest Manierre as 42nd BS commander in mid-1942 but lost his life on 1 February 1943 when leading a section of Fortresses as part of a nine-bomber formation attacking shipping in the Shortlands. Hall's section was intercepted by Zeros and three 42nd BS B-17s were shot down including Hall's 41-9151 and 41-9122 *Eager Beavers* (see Profile 35).

On 7 February 1943 the 42nd BS ceased operations and the original surviving crews from the nine bombers which left Hawaii in July 1942 were relieved. On 20 May 1943 the squadron resumed aerial operations from Hawaii, this time with the B-24D Liberator. During its time in the SOPAC theatre the 42nd BS lost six Fortresses in combat and received nine replacements.

The 42nd BS logo in 1942 was based on a pre-war design that portrayed a blue shield with a golden torch, on top of which is a puma crouching on a tree branch. This is illustrated on the top of page 58.

PACIFIC PROFILES

42nd Bombardment Squadron

Profile 35   B-17E 41-9122 *Eager Beavers*

This was a replacement Fortress that arrived at Palikulo on 18 September 1942. It was damaged during a raid over the Shortlands on 27 September 1942. Following repairs, it was one of the bombers lost over the same target during the disastrous 1 February 1943 mission, as described above.

Profile 36   B-17F 41-24446 *Jezabel*

This "F" model was the first assigned to the 42nd BS after its arrival in the South Pacific in August 1942. However, after only weeks of service it was transferred to the 431st BS on 15 September (see Profile 88). It was later used as a transport by Lieutenant General Nathan Twining.

Profile 37   B-17E 41-2420 *Bessie Jap Smasher*

This Fortress received a HAD camouflage scheme, prior to being one of the original nine 42nd BS machines flown to Espiritu Santo in July 1942. On 24 September 1942 it was damaged in combat with Japanese floatplanes. Pilot Lieutenant Charles Norton nursed the crippled bomber back towards Guadalcanal where it ditched off the north-western coast, behind enemy lines. Norton and one crewman made it ashore but were captured and killed. The rest of the crew remain MIA.

Profile 38   B-17E 41-9216 *Alley Oop*

This Fortress was another of the original nine 42nd BS bombers flown to Espiritu Santo in July 1942. Assigned to Captain Kermit Messerschmitt, it had been named and decorated prior to its departure from the US. *Alley Oop* was a pre-war comic strip. Some time after the 42nd BS ceased B-17 operations, this Fortress was reassigned to the 23rd BS in September 1943.

Profile 39   B-17E 41-2428 *Ole Shasta*

This HAD camouflaged Fortress participated in the Battle of Midway after which in August 1942 it was delivered to the 30th BS at Mareeba. In late 1942 it was ferried to Espiritu Santo and reassigned to the 42nd BS as a replacement. On 28 December 1942 First Lieutenant James Harp flew it over the northern Solomons on an armed reconnaissance mission. It reportedly crashed into the sea off Choiseul and is listed as MIA. It was probably shot down by Zeros.

B-17E 41-2420 *Bessie Jap Smasher*, the subject of Profile 37, in a revetment in Hawaii just before its SOPAC delivery flight.

Vincent Trout "VT" Hamlin (centre) poses with B-17E 41-9216 *Alley Oop*, as illustrated in Profile 38, in Hawaii just before the bomber departed for the South Pacific in July 1942. Hamlin was the creator of the Alley Oop comic strip.

B-17E 41-2428 Ole Shasta, as depicted in Profile 39, at Espiritu Santo.

B-17E 41-9216 Alley Oop, the subject of Profile 38, seen with mission markers at Espiritu Santo.

B-17F 41-24520 Fightin' Swede, the subject of Profile 49, seen at Port Moresby around March 1943.

B-17F 41-24381 Panama Hattie, as shown in Profile 43, at Port Moresby. The crewman sitting against the tyre has "Lowry Field, Colorado" titled on his T-shirt.

CHAPTER 11
63rd Bombardment Squadron

The 63rd BS was activated on 15 January 1941 at Langley Field, Virginia, where it was assigned to the 43rd BG. It commenced training operations with the B-18 Bolo and B-25C before also adding B-17s. The 63rd BS arrived in Sydney aboard the liner *Queen Mary* on 28 March 1942. The men marched into town in heavy rain before catching trams to Randwick Racecourse which became their new camp in Australia. On 11 June 1942 they departed Sydney for Charleville where they were quartered in the local racecourse. On 3 August they moved to Torrens Creek airfield. Here the squadron was assigned a dozen Fortresses, with the first five arriving on 12 August and the remaining seven two days later.

The 63rd BS's aircrew undertook training with the 435th BS at Townsville, before the unit was ordered to Mareeba on 20 August 1942. From this location the 63rd BS aircrews conducted their first missions with the 28th, 30th and 93rd BS. The squadron flew the first combat mission of its own on 12 November 1942, adopting a healthy number of former 19th BG Fortresses. The squadron's bombers continued hitting New Guinea targets from Australian bases until January 1943 when it moved operations to Port Moresby and experimented with skip bombing techniques. In March 1943 the unit was a major participant in the Battle of the Bismarck Sea, striking ships in a Japanese convoy bound for Lae. On 30 June 1943 the 63rd BS lost its first Fortress to a nightfighter when J1N1 Irving pilot FPO2c Kudo Shigetoshi shot down B-17F *I Dood It* flown by First Lieutenant Harold Barnett.

Major William "Bill" Benn took the 63rd BS to war, however, on 13 November 1942 he became the Fifth Bomber Command Chief of Operations and was replaced by Major Ed Scott. The squadron lost nine Fortresses to combat in the New Guinea theatre, with the last loss occurring on 7 July 1943.

Throughout October 1943 the 63rd BS converted to B-24D Liberators which were equipped with low altitude bombing radar with which to conduct "snooper" missions. The squadron then became nicknamed the Seahawks and adopted the motto *Semper Primus* meaning "Always First". Prior to that an unofficial squadron logo had been created in Australia in early 1943. This depicted a castle with wings (hence a "flying fortress") contained in a dropping bomb, as shown at the top of page 64.

The nose artwork on the starboard side of B-17E 41-2609 Loose Goose, as illustrated in Profile 40.

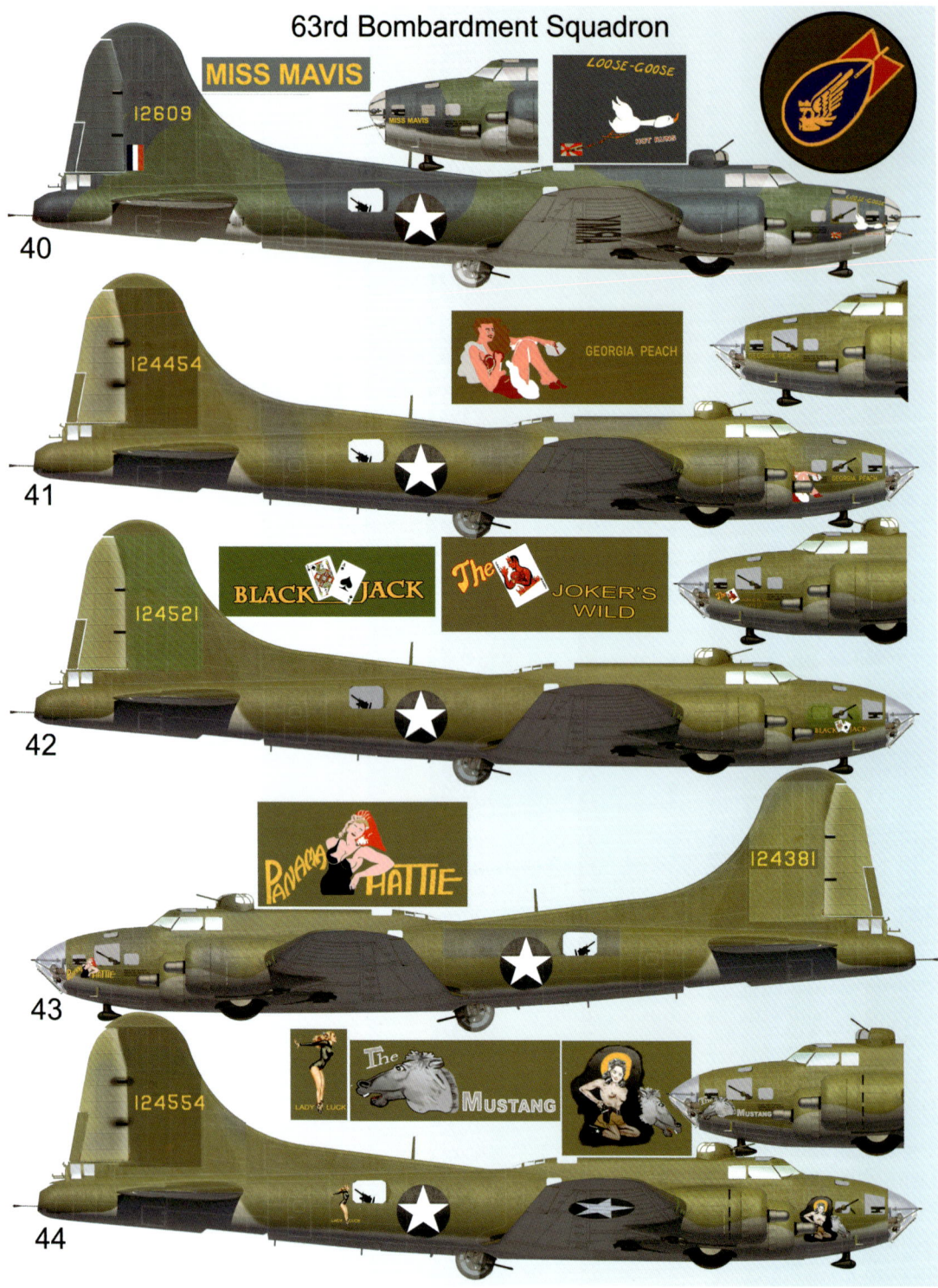

Profile 40   B-17E serial 41-2609 *Miss Mavis / Loose Goose*

Similar to Profile 51, this B-17E was originally intended for RAF Coastal Command and was painted in the relevant camouflage scheme. It was delivered to Australia in April 1942 where it first served with the 19th BG as *Miss Mavis*. When the 63rd BS took delivery of the bomber they ferried it to Mareeba during which a goose escaped from a cage in the back and a new nickname was born. On 20 September 1943 *Loose Goose* participated in one of the squadron's last B-17 missions before the unit converted to the B-24 Liberator.

Profile 41   B-17F serial 41-24454 *Georgia Peach*

This bomber was received from the 19th BG in December 1942 at Mareeba. Its new crew chief Master Sergeant Layton Bacon named it after his wife and had the name painted on both sides of the fuselage. On 13 June 1943 it was caught in searchlight beams and then intercepted from below by a J1N1 Irving night fighter during a night mission over Vunakanau. Critically damaged, it crashed near a hillside on New Britain's northwest coast. Of two men that baled out, First Lieutenant Jack Wisener was transported to a POW camp in Japan and survived the war.

Profile 42   B-17F serial 41-24521 *Black Jack / The Joker's Wild*

This bomber was received as a new airframe by the 63rd BS on 12 September 1942. The bomber was assigned to Captain Kenneth McCullar and early on the ground crew applied the name *Black Jack* on the starboard fuselage after the last two digits of the serial number, "21". Crewmember Technical Sergeant Ernest Vandal painted the name *The Joker's Wild* on the other side. On 11 July 1943 the bomber was ditched off Cape Vogel just south of Finschhafen, following a mission over Rabaul. The crew were rescued, and today the bomber lies on a reef at 45 metres depth and has become a dive site.

Profile 43   B-17F serial 41-24381 *Panama Hattie*

This was one of the first "F" models to arrive in Australia in August 1942. It was used briefly by the 19th BG, prior to assignment to the 63rd BS where it saw much service including participation in the Battle of the Bismarck Sea. In November 1943 it was transferred to the 54th TCW where it was modified to a transport and renamed *Well Goddam* (scc Profile 94).

Profile 44   B-17F serial 41-24554 *The Mustang*

After initial service with the 435th BS from October 1942, this bomber was serving with the 63rd BS by late December 1942. Named *The Mustang*, the nose art was painted by artist and crew chief Technical Sergeant Ernest Vandal. Secondary art was painted behind the starboard waist gunner's position of a brunette named *Lady Luck* stretching forward. The bomber flew in some key 63rd BS missions before in late 1943 it was transferred to the Thirteen Air Force depot in New Caledonia where a decorative 129-mission scoreboard was painted for publicity purposes, not reflecting the bomber's actual combat record. Shortly afterwards the bomber returned to the US where it served as a trainer.

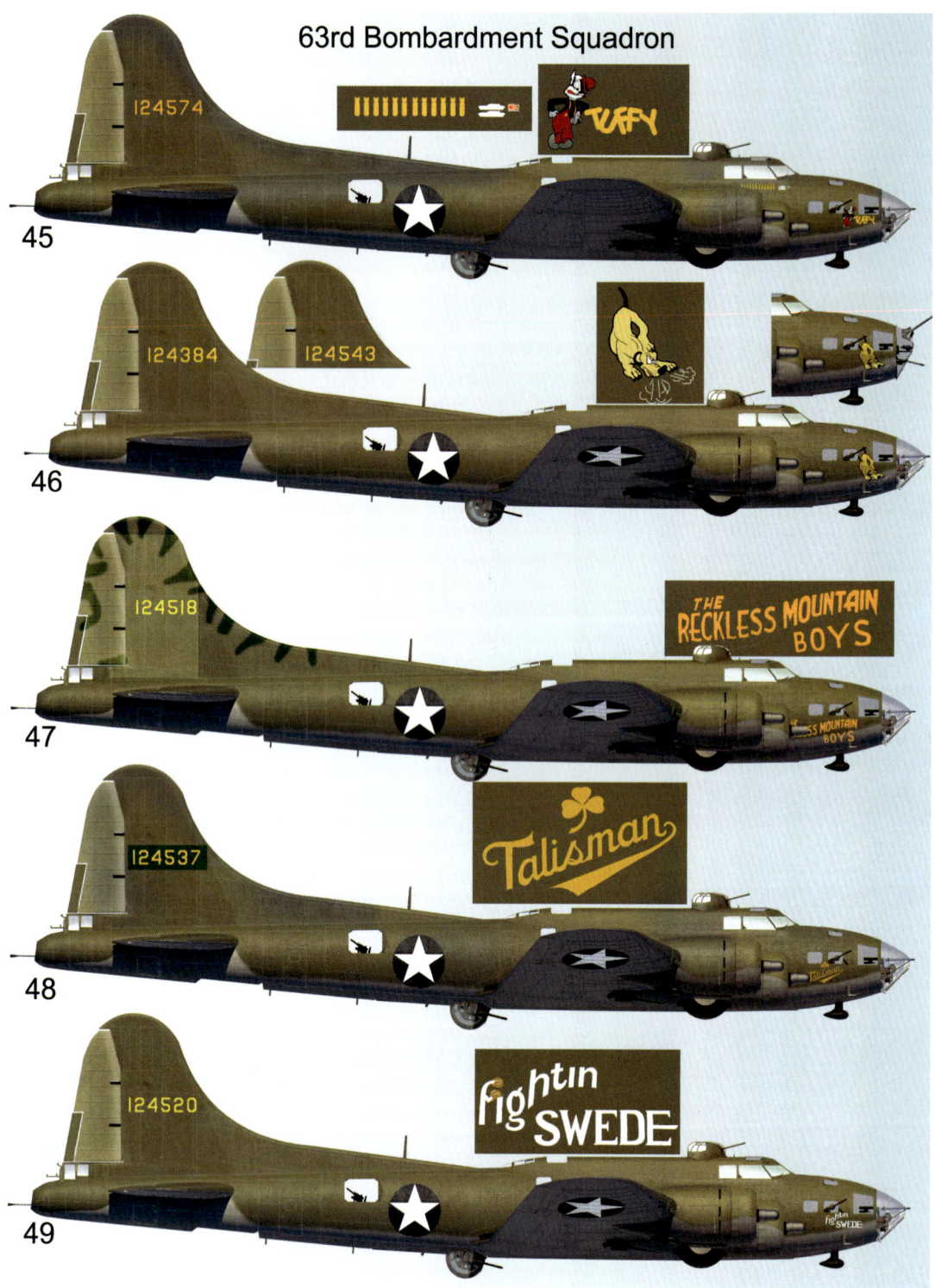

Profile 45   B-17F serial 41-24574 *Tuffy*

This Fortress was delivered to the 63rd BS on 14 November 1942 where it was named *Tuffy* after the cat of the same name which first appeared in the 1942 *Tom and Jerry* cartoon series. After a year of service including extensive combat over New Guinea the bomber returned to the US.

Profile 46   B-17F serials 41-24384 and 41-24543 *Pluto*

These are two separate Fortresses with almost identical nose art. Both bombers were known as *Pluto* however neither had the name applied to the airframe. In the US, the navigator of 41-24384, Second Lieutenant Nathan "Joe" Hirsh, named his bomber *Pluto* and applied the nose-art himself, a direct copy of the Disney namesake character. The bomber arrived in Australia in August 1942 and served with both the 93rd BS and 63rd BS. On 26 March 1943 it departed Port Moresby for an administrative flight to Merauke, carrying several senior Fifth Air Force officers as passengers. However, the bomber disappeared, and its fate is unknown.

Meanwhile Fortress 41-24543 was first assigned to the 403rd BS where it was named *I Dood It*, before transfer to the 63rd BS in February 1943. Shortly after the loss of 41-24384, *I Dood It* was painted over, and a near-identical *Pluto* was added. In the early hours of 30 June 1943, the bomber was shot down by a J1N1 Irving night fighter over Vunakanau.

Profile 47   B-17F serial 41-24518 *The Reckless Mountain Boys*

On 7 September 1942 this bomber was assigned to the 403rd BS which named it *The Reckless Mountain Boys* after a contemporary folk song of the times. The tail was later camouflaged with dark green stripes as illustrated. In early February 1943 the bomber was reassigned into the 63rd BS. On 7 May 1943 it was attacked by Zeros while flying a photographic reconnaissance over Kavieng. Eight crewmen survived a ditching nearby but were captured. Three survived the war as POWs in Japan.

Profile 48   B-17F serial 41-24537 *Talisman*

This bomber was ferried to Australia by First Lieutenant William O'Brien in August 1942 and was assigned to the 63rd BS soon afterwards. The crew named the bomber *Talisman* and decorated it with a three-leafed clover acknowledging O'Brien's Irish ancestry. After an accident on 24 April 1943 the dark green background square to the serial number on the fin was applied during repairs. In September 1943 it was retired from combat. After being stripped to natural metal finish and given a replacement "E" model nose, it served as a transport before it was scrapped in the Philippines in 1945.

Profile 49   B-17F serial 41-24520 *Fightin' Swede*

Handed over to the 63rd BS on 7 September 1942, this Fortress was assigned to First Lieutenant Folmer Sogaard and was named accordingly to reflect Sogaard's Swedish ancestry. It was lost on 8 May 1943 during an early morning armed reconnaissance mission over Madang and Saidor when it was rammed by a Ki-43-I. Note the two boxing gloves on the letter "F".

The port side of B-17F 41-24554 The Mustang, as depicted in Profile 44, at Port Moresby.

The artwork on the starboard side of B-17F 41-24521 Black Jack, as illustrated in Profile 42, seen at Port Moresby. The port side featured The Joker's Wild artwork (see page 69).

The artwork on the port side of B-17F 41-24521 The Joker's Wild, as illustrated in Profile 42.

The starboard side of B-17F 41-24518 The Reckless Mountain Boys, the subject of Profile 47, in a revetment at Seven-Mile 'drome.

B-17E 41-2665 Salisbury's Silly Saps, as depicted in Profile 52, undergoes maintenance at Mareeba.

B-17F 41-24425 "Flagship Oregon", the subject of Profile 54, as it appeared at Mareeba with the 30th BS around September 1942.

CHAPTER 12
64th Bombardment Squadron

A component of the 43rd BG, the 64th BS arrived at Sydney on 15 March 1942, before moving to Daly Waters two months later on 16 May. From 2 August until 12 October 1942, it was stationed at Iron Range, before moving to Fenton where it stayed only a few weeks before moving to Mareeba in November. It was then stationed at Seven-Mile 'drome, Port Moresby, from January until November 1943 during which time the squadron saw heavy combat in New Guinea. After that period the 64th BS transitioned to Liberators. The squadron lost thirteen Fortresses to accidents and combat in the SWPA theatre.

During its B-17 era, the 64th BS incurred a disproportionate loss of squadron commanders in combat. Captain Eugene Halliwill took the squadron to Australia, and was replaced by Major Jack Bleasedale on 15 July 1942. Major Allen Lindberg then assumed command from Bleasedale on 11 December 1942, however, Lindberg was lost over Rabaul on 5 January 1943 when piloting 65th BS B-17F *San Antonio Rose*, along with Bleasedale and Brigadier General Kenneth Walker who were aboard as observers. Walker was the highest-ranking USAAF officer lost in the theatre and the B-17F remains missing to this day. Major Harold Hastings was appointed as the new squadron commander on 15 April 1943 followed by Captain Paul Williams on 15 May 1943, however, Williams in turn was lost when piloting B-17E 41-9244 *Honikuu Okole* (see Profile 51) which was shot down over Rabaul on 21 May 1943. He was replaced by Major Ealon Hocutt the following day, who was in turn replaced by Major Harold Brecht on 7 August 1943 who saw out the 64th BS's B-17 era before it converted to the B-24 Liberator.

This unofficial 64th BS logo is shown in the middle of the profiles on page 72. It was created in Australia and depicted a Red Indian astride a bomb peering through a telescope.

The scoreboard on the nose of B-17E 41-9244 Honikuu Okole, as illustrated in Profile 51. The bomber was shot down in May 1943 and is shown shortly before that in a revetment at Seven-Mile, Port Moresby.

Profile 50   B-17E serial 41-2662 *Spawn of Hell*

This "E" model first served with the 30th BS in mid-1942 and was named *Spawn of Hell*. On the port side a nude baby demon with horns holding a Thompson submachine gun was painted. On 7 November 1942 the bomber was transferred to the 64th BS, but on 31 May 1943 it was transferred to the 65th BS. Some months later it was converted into an armed transport and saw service with the 375th TCG mainly from Port Moresby and Nadzab. It was scrapped in the US in 1946.

Profile 51   B-17E serial 41-9244 *Honikuu Okole*

This B-17E was originally intended for RAF Coastal Command and was painted in the relevant RAF camouflage scheme. In early August 1942 during its trans-Pacific ferry flight at Hickam Field it was named *Honikuu Okole*, a phrase in the Hawaiian language which roughly translates as "Kiss My Ass". First used by the 30th BS, it was received by the 64th BS on 14 November 1942. By the time of its last mission the nose scoreboard showcased seven ships, eighteen fighters and 68 bombing missions, however, the bomber is illustrated as it appeared at Seven-Mile around February 1943. The maingear wheel hubs were decorated with a US insignia as illustrated. On 21 May 1943 it was shot down by a J1N1 Irving night fighter over Vunakanau.

Profile 52   B-17E serial 41-2665 *Salisbury's Silly Saps*

This bomber was first assigned into the 93rd BS in August 1942. It was then reassigned to the 64th BS and pilot First Lieutenant Stanley Salisbury who named the bomber accordingly. It was renamed *Lulu* around July 1943 when Salisbury returned to the US, and then in August 1943 it was reassigned to the 63rd BS. In November 1943 it became one of a dozen Fortresses converted to an armed transport. It was assigned to the 40th TCS where it was renamed *Pretty Baby* (see Profile 103).

Profile 53   B-17F serial 41-24420 *Caroline*

This B-17F arrived in Australia in late September 1942 where it was first assigned to the 28th BS and then to the 65th BS. In January 1943 it was transferred again to the 64th BS with which it flew combat missions until late April 1943 when it was transferred to the 403rd BS. In early November 1943 it was one of the dozen B-17s converted into an armed transport. It was reassigned to the 58th TCS which named it *G.I. Jr.* In 1944 it was ferried back to the US, before it was scrapped in 1946. The bomber is profiled as it appeared with the 64th BS at Port Moresby around February 1943.

Profile 54   B-17F serial 41-24425 *"Flagship Oregon"*

This "F" model was assigned to the 30th BS in August 1942 which named it *"Flagship Oregon"*. It was reassigned to the 64th BS in November 1942. On 17 April 1943 it departed Seven-Mile 'drome flown by Captain Charles McArthur who was then acting squadron commander. Aboard the bomber was an *ad hoc* crew on a formation flight during which it was also intended to calibrate the bomber's compass. While flying in close formation at 1,500 feet the bomber's propellers impacted the tail of B-17F *"Dinah Might?"*. One of the engines on *"Flagship Oregon"* broke from its mounts and caught fire, rendering the bomber uncontrollable. It dived steeply and drove into the sea and the entire crew was listed as MIA that same day. Meanwhile *"Dinah Might?"* managed to return safely to Seven-Mile.

The nose artwork on B-17E 41-9193 Gypsy Rose, as illustrated in Prolife 57.

Artwork and mission markers on B-17F 41-24552 Listen Here Tojo! are clearly displayed in this airborne shot. The bomber is the subject of Profile 56.

CHAPTER 13
65th Bombardment Squadron

Part of the 43rd BG, the 65th BS was formed on 20 November 1940 and the men soon nicknamed themselves *The Lucky Dicers*. Equipped with B-17Es, on 29 August 1941 it moved to Bangor, Maine, where it remained until February 1942 before departing for Australia with the rest of the 43rd BG. After arriving in Sydney on 28 March 1942 it moved to Williamston, New South Wales, on 23 June. Then a series of moves followed to Torrens Creek on 15 August, Iron Range on 13 October, Mareeba on 7 November and then Seven-Mile at Port Moresby on 20 January 1943. Later that year, in December 1943, the 65th BS moved to Dobodura where its B-17 era came to a close.

Squadron commander Major John Roberts oversaw the 65th BS's introduction to combat in the SWPA. He was replaced by Major Harry Hawthorne on 14 January 1943, who was in turn replaced by Captain William Smith on 24 May 1943. Only a month later on 26 June 1943 Major Daniel Cromer took over, who saw out the B-17 era. The 65th BS lost a dozen Fortresses to accidents and combat while in the SWPA.

The 65th BS logo was a pair of tilted dice showing a four and three on the top for the 43rd BG, and a six and five on the front faces representing the 65th BS. This is shown at the top of page 76.

B-17E 41-2634 Red Moose Express, as depicted in Prolife 58, at Mareeba.

Profile 55   B-17F serial 41-24458 *San Antonio Rose*

The most senior Fifth Air Force officer ever lost in combat was aboard this bomber on 5 January 1943, Brigadier General Kenneth Walker, Commander of Fifth Bomber Command. The Fortress was shot down east of Vunakanau Airfield at about 5,000 feet by 11th *Hiko Sentai* Ki-43-Is. Two crewmen baled out and became POWs, but the fate of the rest of those aboard is unknown and they remain MIA.

Profile 56   B-17F serial 41-24552 *Listen Here Tojo!*

This B-17F flew its first 65th BS mission on 22 November 1942 when it bombed Lae. It was lost about a year later on 15 September 1943 when returning to Port Moresby after another bombing mission against Lae. It was in a small formation of Fortresses which had to climb to 11,500 feet to cross the mountains when it entered a thunderstorm about ten miles southwest of Wau. When the others exited the storm the Fortress had gone and was not seen again. It was the last B-17 lost on a combat mission in the SWPA and had in fact crashed into a mountainside at an elevation of 8,300 feet near Black Cat Pass. On the day it was lost the bomber had 48 bomb mission markers, however, it is profiled just after it received its nose art at Seven-Mile before it had flown any missions.

Profile 57   B-17E serial 41-9193 *Gypsy Rose*

This Fortress was assigned to the 65th BS around November 1942. It was named *Gypsy Rose*, a reference to Gypsy Rose Lee, a contemporary striptease performer. The name, written in red and outlined in orange, is reminiscent of brightly lit neon signs of the times. The cursive script used in the lettering of the name is of the style seen on many early 65th BS aircraft. The bomber was lost on 24 May 1943, when it ran out of fuel returning from Rabaul and ditched near Buna. All of the crewmembers were rescued.

Profile 58   B-17E serial 41-2634 *Red Moose Express*

This bomber was transferred to the 65th BS from the 435th BS where it had been named *Tex* (see Profile 33). The new name was a reference to when the 65th BS was stationed in Bangor, Maine, where it had become known as the *Moose Squadron*. On 3 August 1943 it departed Port Moresby on a bombing mission against a Japanese construction camp near Bogadjim. However, it was intercepted and badly damaged over Astrolabe Bay by 24th *Hiko Sentai* Ki-43 Oscars. The bomber entered a steep dive with both inboard engines smoking and broke up when it hit the sea. All of the crew remain MIA.

Profile 59   B-17E serial 41-2430 *Naughty But Nice*

Having flown into Hawaii during the Pearl Harbor raid, this bomber subsequently received the HAD camouflage scheme. After arriving in Australia in February 1942 it was assigned to the 28th BS, and the lower Bendix turret was soon replaced with a Sperry ball turret. In November 1942 it was reassigned to the 65th BS and following numerous combat missions it was lost to a night fighter over Rabaul on 26 June 1943. After its port wing and engines caught on fire the bomber entered a spin and crashed into the Baining Mountains. All aboard were killed except for the navigator First Lieutenant Jose Holguin who survived the war in Rabaul as a POW. The nose art of this bomber is today on display at the Kokopo War Museum near Rabaul.

B-17E 41-9223 *Bitch Kitty*, as illustrated in Profile 60, at Espiritu Santo.

B-17E serial 41-9156 *Uncle Biff*, as depicted in Profile 64, cruises over the Solomons.

CHAPTER 14
72nd Bombardment Squadron

Part of the 5th BG, the 72nd BS was the first squadron from that group to deploy to the South Pacific, when on 15 September 1942 eight B-17Es under Major Don Riddingstook departed from Bellows Field, Hawaii. Staging via Fiji, they arrived at Espiritu Santo on 24 September. Meanwhile other B-17s in the theatre were transferred into the squadron to bolster it to full strength. Key maintenance personnel were ferried ahead of the flight echelon from Oahu in LB-30s, the transport version of the Liberator. From Espiritu Santo the B-17s commenced a series of long-range patrols, extending as far as 800 miles. From October 1942 the 72nd BS was integrated with the 11th BG for several weeks.

From late 1942 the squadron mostly operated from Guadalcanal. Then at the end of February 1943 its entire B-17 inventory was transferred to the 31st and 23rd Bombardment Squadrons, as the B-17s were replaced by B-24 Liberators. The 72nd BS lost eight B-17s in the SOPAC theatre to accidents and combat.

The 72nd BS logo derives from 1924. It portrays "72" as stylised lightning bolts in front of thunderclouds, and is shown at the top of page 80.

B-17E 41-9059 Boomerang, the subject of Profile 61, at Henderson Field.

Profile 60   B-17E serial 41-9223 *Bitch Kitty*

This bomber was one of the original eight 72nd BS machines which arrived at Espiritu Santo on 24 September 1942. It was assigned to Captain James "Ted" Joham from Santa Barbara, California, who first flew from Guadalcanal on 4 October for attacks on Buka. He flew again the following morning against Gizo and Rekata Bay. The bomber was named at Espiritu Santo around this time but was written off on 4 November 1942 following an accident.

Profile 61   B-17E serial 41-9059 *Boomerang*

This bomber arrived in the New Hebrides in July 1942 and initially flew with another squadron. It was reassigned to the 72nd BS on 18 September 1942 just before the rest of the unit arrived from Hawaii. At Espiritu Santo it was fitted with a SCR-521 ASV search radar. During a mission to Rabaul on 25 December 1942, the outer starboard engine failed and the pilot aborted the mission. Thirty minutes later during the return to Guadalcanal the #2 engine failed followed by the loss of the #1 engine. Incredibly the bomber made it back safely to Henderson Field with just ten minutes of fuel remaining. With a bad reputation within the squadron, this was the bomber's final mission after which it was scrapped.

Profile 62   B-17E serial 41-2523 *Goonie*

Ferried from Hawaii to Plaine De Gaiacs in July 1942 with the 98th BS, by 1943 this Fortress had been reassigned to the 72nd BS. On 20 March 1943 it departed Guadalcanal flown by Lieutenant Colonel Marion Unruh on a night mission against Buin airfield. The Fortress was hit by an anti-aircraft shell burst which caused a runaway engine which Unruh was unable to feather. He successfully ditched the bomber off the Russell Islands, and the crew were all rescued and returned to duty.

Profile 63   B-17E serial 41-9093 *Spoook! / Miss Betty*

This Fortress was assigned to the 72nd BS in Hawaii on 7 June 1942. Subsequently at various times in the South Pacific it also served with the 431st and 31st Bombardment Squadrons, where it was named *Miss Betty*. Finally, it was transferred again to the 23rd BS on 15 June 1943, before it returned to the US six months later.

Profile 64   B-17E serial 41-9156 *Uncle Biff*

After serving with the 72nd BS on Hawaii, this bomber was ferried to the South Pacific by the 431st BS in July 1942. On 29 September 1942 it was attacked by Zeros over the Shortlands and returned to Henderson Field with major bullet and shrapnel damage, although somehow no-one aboard was injured. In January 1943 it was transferred to the 31st BS just before it was ferried back to Hawaii for a major refit. In July 1943 it was reassigned to the 394th BS, then again to the 23rd BS. It was retired from combat duty in September 1943, and finally scrapped in June 1944.

B-17E 41-2632 Crock o' Crap, the subject of Profile 65, at Nadi, Fiji, during the November 1942 return to Hawaii. Mt Batilamu is in the distance to the far right.

B-17E 41-2489 Suzy-Q, as illustrated in Profile 66, during its war bond tour of the US.

CHAPTER 15
93rd Bombardment Squadron

Part of the 19th BG, the 93rd BS was originally deployed to Clark Field in the Philippines in October 1941. Then on 6 December the squadron relocated to Del Monte, a new field established in Mindanao as a dispersal measure. From here they engaged in combat from secondary airfields against the invading Japanese forces until the situation in the Philippines became untenable and they were withdrawn to Australia. The escaped airmen and Fortresses reformed and became involved in the New Guinea campaign flying missions mainly from Longreach in Queensland and staging through Port Moresby. Some of the earlier Rabaul missions returned to Horn Island rather than Port Moresby, to avoid air-raids, followed by an almost six-hour flight back to Longreach.

Unlike the 19th BG's other squadrons and following the arrival of new "F" models in the theatre in September 1942, on 25 October 1942 the 93rd BS ferried its dozen surviving B-17Es back to the US via New Caledonia and Fiji. These were serials 41-2438, -2440, -2453, -2486, -2489, -2630, -2632, -2642, -2644, -2658, -2668 and -2669, seven of which were held over, overhauled and reassigned to the 394th BS at Hawaii. Many then served in the South Pacific theatre. Between June and September 1942, the 93rd BS lost five bombers to accidents, and three more in combat.

The 93rd BS logo was a 1924 pre-war design which depicts a yelling Indian in war paint with two feathers and a necklace of teeth. On some designs the number "93" was superimposed over the feathers. Although left facing, the logo always faced forward when applied to an airframe meaning it was reversed when applied to the starboard side. It is illustrated at the top of page 84.

B-17E 41-2440 *Calamity Jane*, as depicted in Profile 68, at Nadi in mid-November 1942 following its ground collision with a VS-6 SBD.

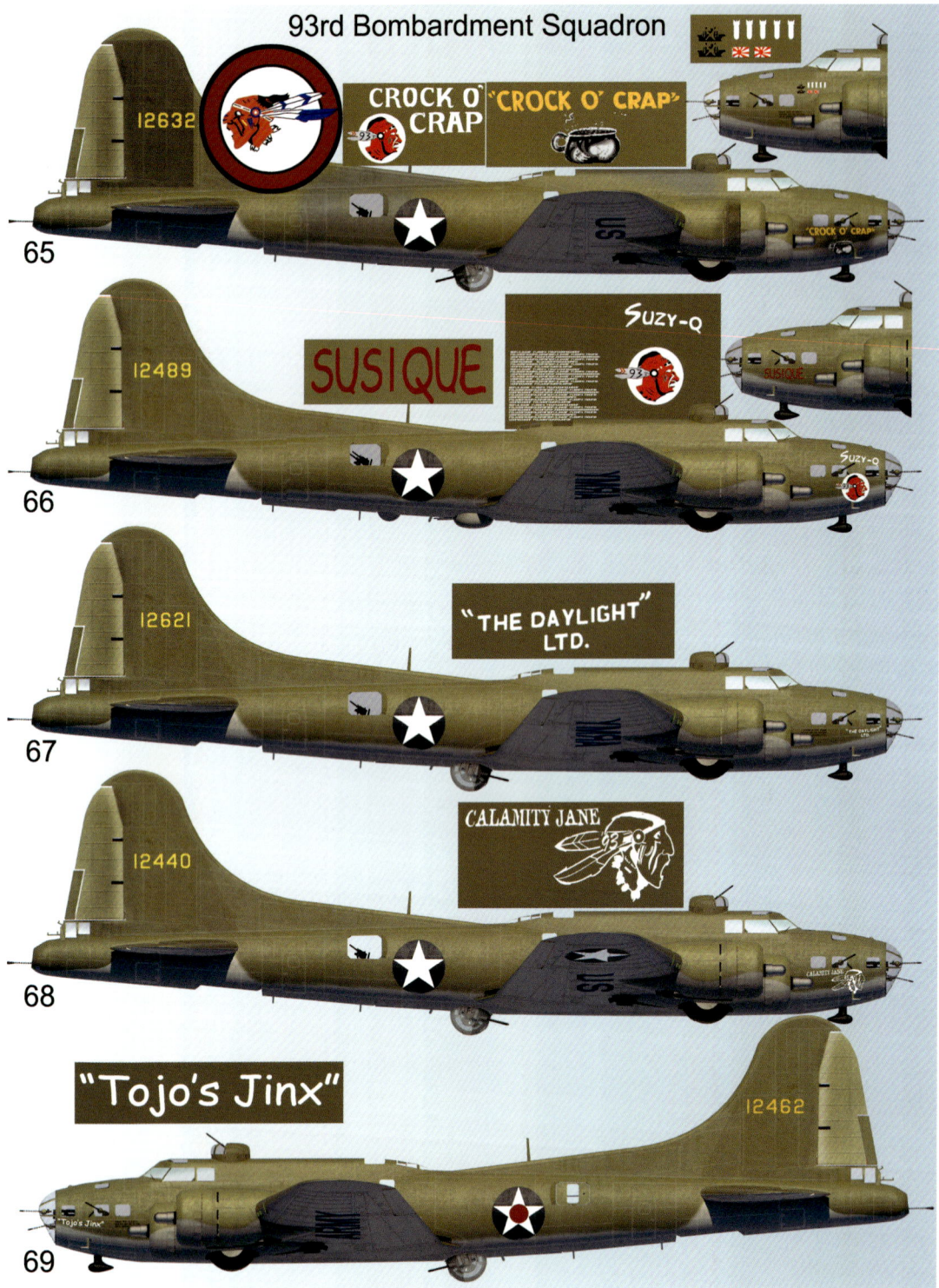

Profile 65   B-17E serial 41-2632 *Crock o' Crap*

This Fortress was sent to Australia in July 1942 to bolster the 93rd BS. In November 1942 it was among the dozen surviving B-17Es ferried back to Hawaii for overhaul. Transferred to the 394th BS, it arrived at Espiritu Santo on 3 January 1943. During the year it operated over the Solomons before returning to the US on 30 December 1943. Its first form of art employed the 93rd BS logo, whereas the second replaced this with more graphic art. The scoreboard is illustrated as it appeared at the end of its tour, *en route* back to Hawaii.

Profile 66   B-17E serial 41-2489 *SUSIQUE / Suzy-Q*

This bomber became the most famous in the 93rd BS as it was employed to do a war bond drive across the US after its Pacific tour. It was first assigned to squadron commander Major Felix Hardison who named it after his wife after it was assigned into the squadron on 20 February 1942 at Longreach. It returned to the US in November 1942 where it was used for a war bonds tour. Its first name of *SUSIQUE* appeared in red letters on the port side. This was later supplemented by the 93rd BS Indian Head insignia on the other side and the name *Suzi-Q*.

Profile 67   B-17E serial 41-2621 *The Daylight Ltd*

This bomber was written off in a crash landing at Mareeba on 26 August 1942 after being hit by anti-aircraft fire while bombing Japanese ships off Milne Bay. It had only been in the 93rd BS since 8 July 1942 when it arrived in Australia after being ferried from the US. The pilot found it difficult to control the bomber during the return flight and circled Mareeba airfield before landing. Without brakes or flaps one maingear sheared off causing the bomber to ground loop to starboard where it impacted a tree.

Profile 68   B-17E serial 41-2440 *Calamity Jane*

This Fortress was among the first to arrive in Australia with the 14th RS which then became the 435th BS. It was combat damaged in July 1942 and following repairs it was reassigned into the 93rd BS at Mareeba. In October the bomber was one of those sent back to the US where it was transferred to the incoming 394th BS. However, on its way to the South Pacific it was damaged in a ground collision with an SBD in Fiji. Meanwhile the bomber had acquired the nickname *Calamity Jane*, and the name stuck so it was painted on the nose, however, over the following weeks the artwork took on several forms by the addition of a nude pin-up girl (see Profile 77). It is illustrated in this profile as it appeared in Fiji just after the first version of its nose-art was applied. After local repairs the bomber was ferried back to Hawaii for further overhaul before being returned to the 394th BS. Further history on this bomber can be found in Profile 77.

Profile 69   B-17E 41-2462 *"Tojo's Jinx"*

This bomber arrived in Australia on 20 February 1942 and flew numerous combat missions with the 93rd BS throughout New Guinea. Following damage from strafing at Seven-Mile in December 1942 it was repaired and then refurbished at the USAAF 4th Air Depot at Garbutt into transport configuration. It then became the personal transport of Lieutenant General Walter Krueger, the commander of the US Sixth Army (see Profile 97).

B-17E 41-2621 The Daylight Ltd, as depicted in Profile 67, at Mareeba on the afternoon of 26 August 1942 following its crash landing.

B-17E 41-2658 at Seven-Mile during a brief New Guinea deployment decorated with the 93rd BS insignia, a common marking of the times. This bomber appears in Profile 75 when it later served with the 394th BS.

The second-generation nose art on B-17E 41-2632 Crock o' Crap, the subject of Profile 65, at Henderson Field in early 1943 when the bomber was serving with the 394th BS.

B-17E 41-2462 Tojo's Jinx prepares for a night departure from Longreach to stage through Port Moresby to bomb Rabaul. Several early Rabaul missions returned to Horn Island rather than Port Moresby, to avoid air raids, incurring a six-hour return flight to Longreach. The bomber is illustrated in Profile 69.

B-17E 41-9215 Galloping Gus, the subject of Profile 73, as it came to rest at Palikulo on 18 November 1942.

The artwork on B-17E 41-9219 Typhoon McGoon, as illustrated in Profile 71.

CHAPTER 16
98th Bombardment Squadron

The 98th BS was activated on 16 December 1941 within the parent 11th BG at Hickam Field, Hawaii. From 21 July 1942 this squadron and its counterpart 42nd BS began arriving at Plaine des Gaiacs. The 98th BS's original nine B-17Es were based at Plaine des Gaiacs, a rudimentary airfield which then featured New Caledonia's longest runway of 7,000 feet. However, only nine revetments were large enough to accommodate bombers, so to avoid congestion the 42nd BS B-17Es were briefly detached to Koumac whilst the 98th BS remained *in situ*. Both squadrons flew search and reconnaissance missions from these locations up until the invasion of Guadalcanal. For these missions the B-17Es were forced to borrow cameras loaned from VMO-251. These sorties proved of marginal value when most were thwarted by difficult meteorological conditions.

On 11 August 1942 the 98th BS moved to Palikulo on Espiritu Santo. On 12 September 1942 four 98th BS Fortresses plus two more from the 431st BS departed to bomb a reported Japanese task force. The report was wrong, however, and when returning in bad weather and low on fuel, three bombers including B-17E 41-9219 *Typhoon McGoon* (see Profile 71) ditched in early morning darkness on 13 September.

The 98th BS commenced combat operations from Henderson Field in November 1942 but returned to Palikulo on 5 December. On 8 April 1943 the squadron returned to Mokuleia Field on Oahu to commence transition to B-24 Liberators. The 98th BS lost five Fortresses to accidents and combat while in the SOPAC theatre.

The 98th BS logo was designed in the theatre but was not officially approved until 1944. It depicts a falling cartoon-style yellow bomb with a face displaying imprudence, and is illustrated on the top of page 90.

B-17E 41-9211 *Typhoon McGoon II*, as depicted in Profile 72, at Palikulo, Espiritu Santo, in late 1942.

Profile 70   B-17E serial 41-2616 *"Blue Goose"*

The greatest colour scheme myth to evolve from the Pacific War is that of the Fortress "Blue Goose". It stems from an interview with 11th BG commander Colonel Laverne "Blondie" Saunders, during which he described a 1942 mission which he led against Tulagi. During the mission attacking fighters made a beeline for Frank Waskowitz who was Saunders' wingman. Describing the incident Saunders said "Fritz has the only off-color plane in the outfit. It is painted a sort of baby blue and Fritz calls it his blue goose". From this quote stemmed a suite of fictitious and surreal all-over blue schemes for the *"Blue Goose"*.

The B-17E in question was certainly "off-color" compared to its Olive Drab counterparts; in fact it was painted in the RAF Temperate Land and Deep Sky scheme, extending to the fin thus appearing mostly blue when viewed from a level azimuth. While no photos of the bomber have surfaced to date, it was built to RAF specifications with the British serial FK193, before it was reassigned to the USAAF. On 29 September 1942 the bomber was shot down with the loss of all aboard.

Profile 71   B-17E serial 41-9219 *Typhoon McGoon*

This Fortress was assigned into the 98th BS at Hawaii in June 1942. On 12 September 1942 it failed to return from a mission to attack a Japanese task force reported northwest of the Santa Cruz Islands which proved to be a false sighting. In the early hours of 13 September and down to no fuel the bomber was ditched off New Caledonia, with the crew safely reaching the shore using inflatable dinghies.

Profile 72   B-17E serial 41-9211 *Typhoon McGoon II*

This Fortress was assigned into the 98th BS at Hawaii in June 1942, before arriving in New Caledonia the following month. After flying a dozen missions, it was fitted with an SCR-521 ASV search radar at Espiritu Santo in mid-September, and during this refit it was named *Typhoon McGoon II* in honour of the first *Typhoon McGoon* (see Profile 71 above). It wound up serving also with the 23rd and 394th BS, before serving as a transport from late 1943.

Profile 73   B-17E serial 41-9215 *Galloping Gus*

Assigned into the 98th BS at Hawaii in June 1942, similar to other B-17Es in the unit it had an additional 0.50-inch calibre machine gun mounted in the upper radio compartment hatch. It was badly damaged by anti-aircraft fire on 18 November 1942 while over Tonolei harbour. The bomber was written off upon return to Palikulo where it over-ran the runway.

Profile 74   B-17E serial 41-9214 *The Skipper*

Initially assigned to the 98th BS when it moved to Mokuleia Field, Hawaii, in March 1942, *The Skipper* was subsequently transferred to the 23rd BS at Espirito Santo in October 1942. It then operated from Guadalcanal until ferried back to the US in December 1943. The bomber was named *The Skipper* when it was with the 98th BS at Espiritu Santo and continued to carry the name with the 23rd BS.

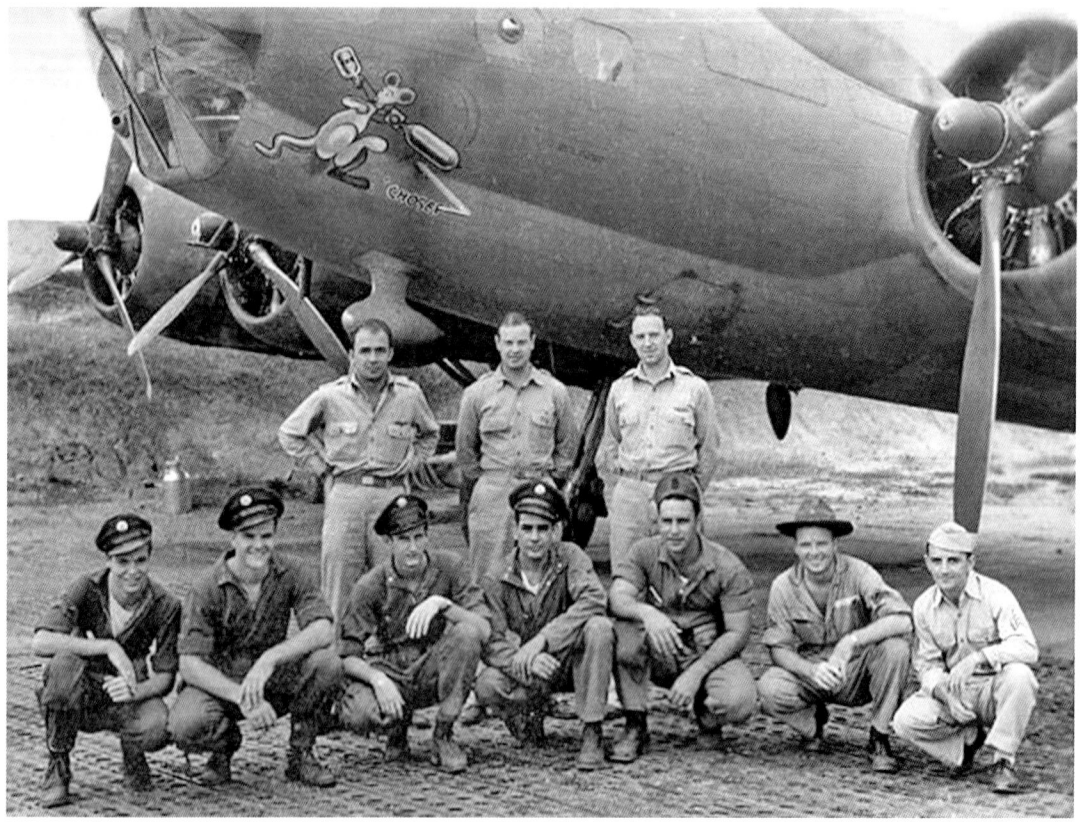

The assigned pilot of B-17E 41-2668 Chosef, Captain John Turner, poses for a crew shot in front of the bomber at Henderson. Chosef is illustrated in Profile 76.

CHAPTER 17
394th Bombardment Squadron

A component of the 5th BG, the 394th BS stayed in Hawaii for most of 1942 from where it flew patrols and trained bomber crews. In early November 1942 it moved to Bellows Field to prepare to move to the South Pacific under the command of Midway veteran Major Orin Rigley. Its initial B-17E inventory comprised seven ex-19th BG B-17E old-timers ferried back from Australia, which had been refurbished in Hawaii. Accordingly, several airframes retained the Indian head insignia of the 93rd BS as illustrated in Profile 75.

On 15 November Rigley led these first seven Fortresses in departing Hawaii for the South Pacific, however, one bomber soon returned after an engine failure and then Captain William Ivey also turned back with a windmilling propeller. The other five continued to Canton Island and then Nadi, Fiji, where they arrived on 18 November, with the two stragglers catching up over the next two days. The 394th BS commenced long-range sector patrols from Nadi usually at a range of 700 miles. Meanwhile the squadron had a shortage of navigators so officers from other disciplines were offered the opportunity to qualify as navigators. At Nadi the unit set up its own *ad hoc* school, dubbed colloquially as the "Royal Fiji Institute of Navigation" whose graduates included several former armaments officers. A 394th BS incumbent who later became famous was pilot Second Lieutenant Gene Roddenberry, the creator of the TV series *Star Trek*. He flew most of his 89 missions as the co-pilot of B-17E *Los Lobos* 41-2644 (see Profile 96).

On 3 January 1943 the 394th BS was ordered to Henderson Field via Espiritu Santo. However, only five B-17Es were able to depart Nadi, and three of these then dropped out with various mechanical problems. Hence only two bombers arrived at Guadalcanal the next day: 41-2632 *Crock o' Crap* (Profile 65) and 41-2630 *Oklahoma Sooner* (Profile 78). The squadron flew missions against Bougainville targets before relocating to Espiritu Santo on 19 January 1943. There they waited for more refurbished B-17Es, whilst also taking time to build an officers' club and take leave in New Zealand.

When Rigley was promoted and transferred to 5th BG Headquarters in April 1943, he was replaced by the recently promoted Major McLyle Zumwalt who led them back to Guadalcanal on 25 April 1943. Missions recommenced on 28 April, however, for the new few weeks they were constantly turned back by frontal systems. Finally on 13 May Zumwalt led nine Fortresses against Buin and Ballale airfields at night, the bombers toting powerful M-26 flares in addition to their bomb load. The idea was that each B-17 would drop by the light of the flares dropped by the plane ahead, then in turn release its own flares to repeat the process. Meanwhile, the run of bad weather continued.

Supporting a USN plan to mine Tonolei Harbour on 20 May, four 394th BS Fortresses provided a diversion using fragmentation and 300-pound bombs, concentrating on searchlight and anti-aircraft positions while TBFs dropped mines from low altitude. The mission was successfully repeated on 23 May, this time with eight B-17s distracting enemy defences. The 394th BS returned to Nadi on 5 June to conduct more long-range patrols, which it finished on 16 June. The squadron then returned to Guadalcanal, this time Carney Field, a new airfield some seventeen miles east of Henderson, for effectively their third combat tour. However, it was here that morale sank to an all-time low. With all the recent turn-backs and maintenance issues the crews felt their luck had finally run out, compounded by ongoing resentment from having been allocated second-hand Fortresses. Meanwhile the Thirteenth Air Force saw an opportunity to restructure its Fortress squadrons, and all the 394th BS flight crews were stood down from active duty, with newcomers transferred to other squadrons. A few days later the veterans boarded a Dutch freighter and arrived at San Francisco on 27 August 1943. The 394th BS was then redesignated a B-24 Liberator outfit and all of its B-17Es were transferred to the 23rd BS. The 394th BS lost only two Fortresses in the SOPAC theatre, both to accidents.

Although modern in appearance, the 394th BS logo was nonetheless created in 1931, and portrays a Cross Etoile split symmetrically between blue and orange, as shown at the top of page 96.

B-17E 41-2440 Calamity Jane, as depicted in Profile 77, at Tontouta just before it was ferried back to Hawaii. The previously topless pin-up girl has received a modest white shirt.

B-17E 41-2630 Oklahoma Sooner, the subject of Profile 78, at Tontouta just before it headed back to the US for the last time.

The artwork on B-17E 41-2642 Poison Ivey, as depicted in Profile 79.

Profile 75   B-17E serial 41-2658

This Fortress was one of the 93rd BS Fortresses ferried back to Hawaii in October 1942. There it was transferred to the 394th BS but was badly damaged while parked in Fiji on 25 November 1942 when a cloud of dust blinded a departing VS-6 SBD pilot, who crashed into several Fortresses. This bomber was written off and is illustrated as it appeared on the day of the accident, still with the 93rd BS Indian logo on the nose.

Profile 76   B-17E serial 41-2668 *Chosef*

This Fortress was named *Chosef* by its assigned pilot Captain John Turner and was decorated with a boxing kangaroo from the *Snuffy Smith* comic strip on the nose. When the 394th BS was disbanded as a Fortress unit the bomber was transferred to the 23rd BS.

Profile 77   B-17E serial 41-2440 *Calamity Jane*

As explained in Profile 68 this Fortress had previously served with the 93rd BS. After transfer to the 394th BS in Hawaii it had been damaged at Nadi on 25 November 1942 when a VS-6 SBD collided with several parked bombers. To make sure it could make the flight to Hawaii for repairs, it underwent local repairs in New Caledonia. There the base commander ordered that a white shirt be painted over the nude upper torso of the artwork, so it would be more acceptable to the more conservative tastes applying far from the frontlines. After overhaul for a second time in Hawaii in December 1942, it served with the 394th BS in the Solomons during 1943 until its final return to the US in March 1944.

Profile 78   B-17E serial 41-2630 *Oklahoma Sooner*

This former 93rd BS Fortress was ferried back to Hawaii in late October 1942 by Captain Bernie Barr. There it was reassigned to the 394th BS and was one of only two of the squadron's bombers that arrived at Henderson Field on 4 January 1943. It was transferred to the 23rd BS in August 1943.

Profile 79   B-17E serial 41-2642 *Poison Ivey*

This former 93rd BS Fortress was named by Captain William "Bill" Ivey in Hawaii just before it flew to the South Pacific, arriving at Nadi on 15 November 1942. After operations from Guadalcanal and Espiritu Santo in 1943 it was ferried back to the US. In March 1944 it was written off after a ground collision with a B-29.

The artwork on B-17E 41-2417 Monkey Bizz-Ness, as depicted in Profile 83, at Seven-Mile in 1943.

Super Snooper (not illustrated) was B-17F 41-24420 which also served with the 403rd BS. It is seen at Seven-Mile in 1943.

CHAPTER 18
403rd Bombardment Squadron

The 403rd BS was originally formed as the 13th Reconnaissance Squadron (Heavy) in 1940 at Langley Field, Virginia. Its first assigned aircraft was a B-18 Bolo with two PT-17s used for flight training. On 15 January 1941 the squadron was transferred into the parent 43rd BG. On 23 August 1941 it moved to Bangor, Maine, from where its cadre embarked on 18 January 1942 aboard the SS *Argentina* bound for Australia. The ship docked at Melbourne on 27 February 1942 where the squadron was initially based at Camp Darley. On 3 March 1942 an advance detachment arrived at RAAF Laverton to prepare an encampment for the rest of the squadron which arrived nine days later. At RAAF Laverton the ground crew commenced refurbishment of war-weary Fortresses on which aircrews commenced flight training.

On 22 April 1942 the unit was redesignated as the 403rd BS under commander Major Thomas Charles, then Major Franklyn Green took over from 5 December 1942. On 27 August 1942, the squadron moved to Torrens Creek in Queensland, and then on 17 October 1942 to Iron Range from where it commenced its first combat missions. On 23 November 1942 the squadron deployed to Turnbull airfield at Milne Bay in New Guinea. On 21 January 1943 it returned to Queensland, this time to Mareeba, led by a new squadron commander, Major Jay Rousek. During March 1943 it participated in the Battle of the Bismarck Sea, and from May until July 1943 it operated from Port Moresby. After this it commenced transition to the B-24 Liberator under another new commander, Major William Welch. During its time in the SWPA theatre the 403rd BS lost four Fortresses to accidents and two to combat.

The 403rd BS used an unofficial "Mareeba Butchers" logo throughout the war which later appeared on the squadron's Liberators. The logo was never approved and was replaced in 1959. Its origins stem back to early 1943 Tokyo Rose broadcasts which called all Fortress units based at Mareeba "butchers", due to their bombing of Rabaul. The identity was readily adopted by the squadron whilst still based at Mareeba, and the logo is shown at the top of page 100.

B-17E serial 41-2481 Topper, the subject of Profile 84, under repair at Mareeba.

403rd Bombardment Squadron

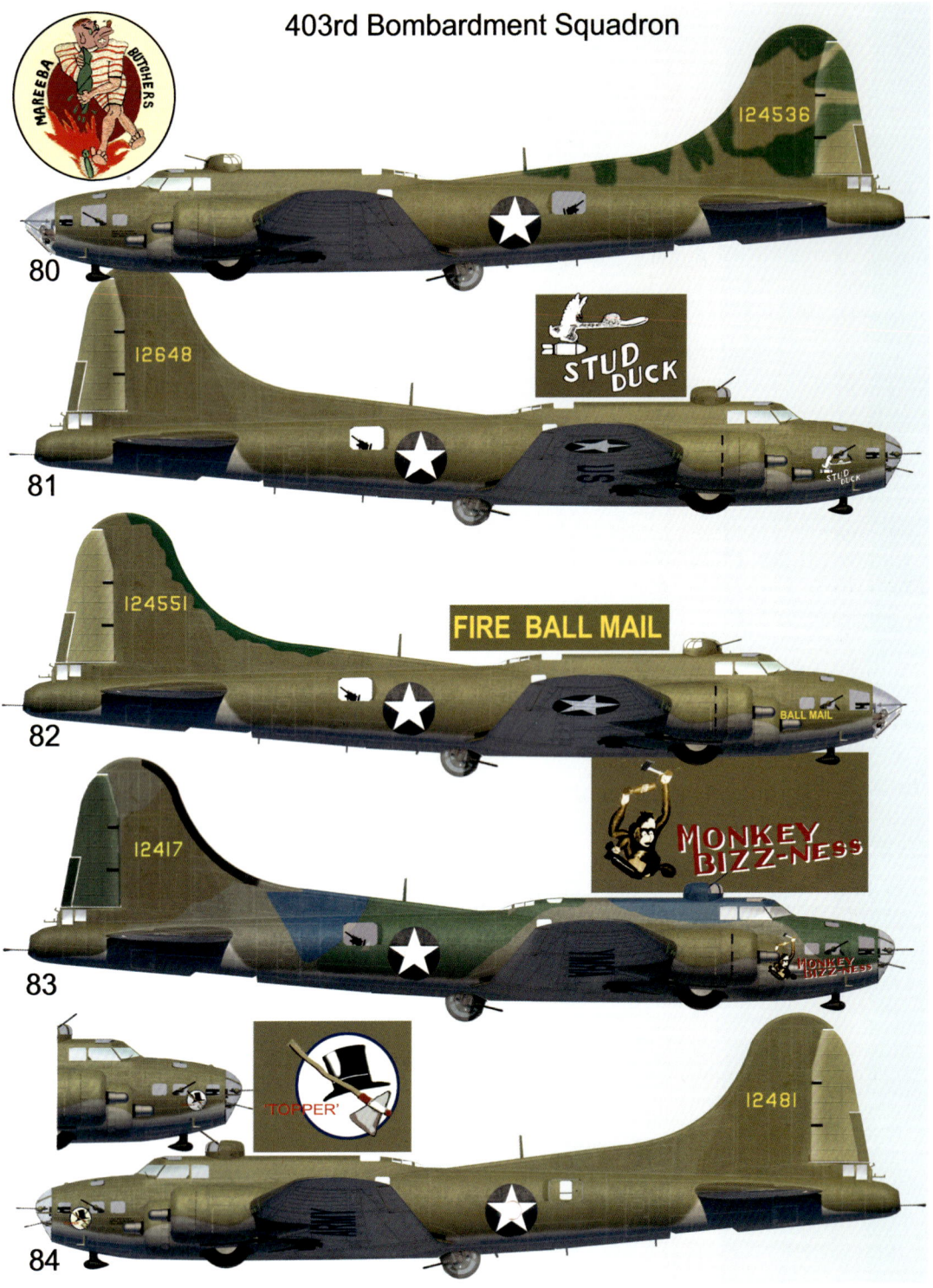

Profile 80   B-17F serial 41-24536

This "F" model was ferried to Australia via Hawaii, arriving on 14 September 1942. It was assigned to the 403rd BS at Torrens Creek, but the following month it was transferred to the 65th BS. Following combat service with a variety of units it was ferried back to the US in June 1944. The unnamed bomber was distinguishable by an unusual dark green camouflage pattern on the fin and is illustrated as it appeared at Port Moresby with the 63rd BS in mid-1943.

Profile 81   B-17E serial 41-2648 *Stud Duck*

A Battle of Midway veteran, this bomber was delivered to Australia shortly thereafter and assigned to the 435th BS. By April 1943 it had been transferred to the 64th BS which named it *Stud Duck*, a colloquialism of the times referring to any big boss heading an organisation. In June 1943 the bomber was reassigned to the 403rd BS. Retaining the name, it was used mainly for transport and reconnaissance missions. Following combat damage and repairs it was reassigned to the 65th BS where it was renamed *Lil Buster Upper*. In October 1943 it was earmarked for a US war bond tour, although this was subsequently cancelled. The bomber is illustrated just after its transfer to the 403rd BS at Port Moresby.

Profile 82   B-17F serial 41-24551 *Fire Ball Mail*

This bomber was first assigned to the 63rd BS at Mareeba on 18 September 1942. Shortly thereafter it was transferred to the 403rd BS on 23 November 1942 which named it *Fire Ball Mail*. The bomber was set afire by bombs in a revetment during an air raid by Betty bombers against Milne Bay's Turnbull Field on 17 January 1943.

Profile 83   B-17E serial 41-2417 *Monkey Bizz-Ness*

Camouflaged in the HAD scheme before it departed Hickam on 6 January 1942, this early "E" model was among the first trio of Fortresses to pave the inaugural ferry flight to Australia. After service in Java, it was reassigned to the 28th BS and then the 63rd BS where crew chief and artist Technical Sergeant Ernie Vandal painted the name *Monkey Bizz-Ness* alongside artwork of a monkey holding a liquor bottle and an axe. Thus named, it entered service with the 403rd BS in February 1943 at Port Moresby and was later converted to a VIP transport (see Profile 99).

Profile 84   B-17E serial 41-2481 *Topper*

Exceptionally, this bomber was ferried to Australia via the Africa Route and the NEI after departing the US on 11 January 1942. Prior to departure the *Topper* artwork was applied by Second Lieutenant Paul Eckley, illustrating a top hat with a tomahawk inside a circle. The name emulated a 1937 comedy starring Cary Grant with the same title. In November 1942 the bomber was reassigned to the 65th BS, then in February 1943 it was reassigned to the 403rd BS. Following numerous combat missions, on 27 August 1943 during take-off from Seven-Mile the starboard main tyre blew out. Heavily loaded with two 1,000-pound and three 500-pound bombs, the blowout swerved the bomber off the runway collapsing the starboard maingear. No-one aboard was injured, however, the bent airframe was considered a write-off.

B-17E 41-2627, as depicted in Profile 86, at Seventeen-Mile 'drome, with the R.F.D. Tojo stork artwork visible on the starboard side of the nose.

The three main types of reconnaissance cameras used in the Pacific theatre, showcased here at Seven-Mile in April 1943. These large cameras required modified fuselage brackets, and the 8th PRS modified its B-17 "camera ships" accordingly, including with the installation of an additional six 0.50-inch calibre machine guns for defensive firepower during solo missions.

CHAPTER 19
8th Photo Reconnaissance Squadron

The 8th PRS was formed at March Field, California, on 1 February 1942 comprising Flights A, B and C. The squadron first operated F-4 and F-5 Lightnings in the SWPA, orchestrated by commander Lieutenant Karl Polifka who initiated a systematic mapping survey of all enemy bases in New Guinea. The Lightnings staged through Port Moresby to achieve this, in addition to surveying potential Allied bases. When the need for long-range assignments became obvious, later from Port Moresby's Seventeen-Mile 'drome the 8th PRS operated a small detachment of B-17Es for long-range photographic reconnaissance. From 16 July 1942 the squadron's first B-17E had been attached to the 19th BG before it was administratively transferred to the 8th PRS in October 1942. Subsequently, the detachment continued to work closely with the 43rd BG in New Guinea in early 1943. The squadron was known colloquially throughout New Guinea as the *Eight-Ballers*.

The 8th PRS logo featured an Indian Brave, who represented the men in the squadron, brandishing a hatchet to signify war with a camera around his neck. In addition, there was a lightning bolt to represent the Lockheed Lightning and an eight-ball. The logo is shown at the top of page 104.

The revised nose art on B-17E 41-2627 Blanche, as shown in Profile 86, at Schwimmer 'drome, Port Moresby.

PACIFIC PROFILES

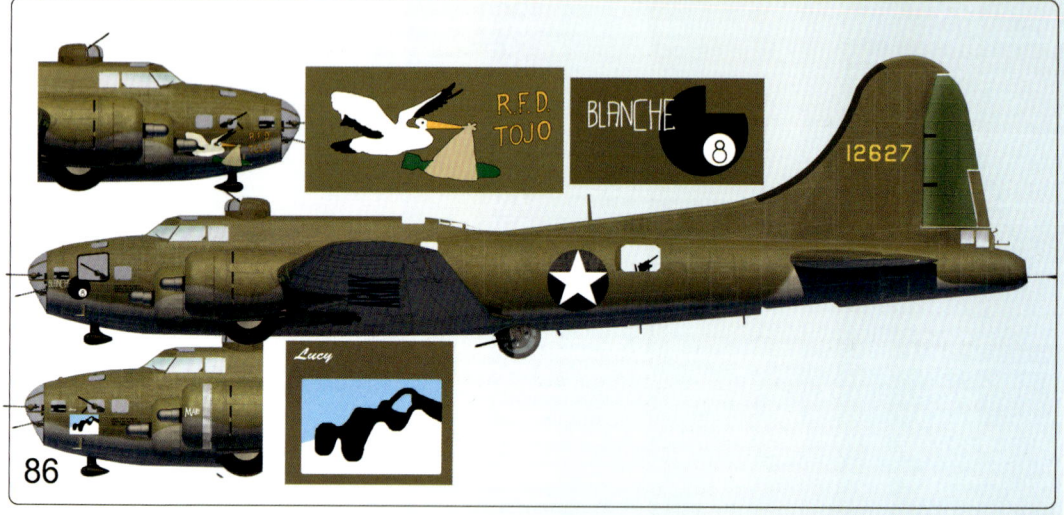

B-17E 41-2627 Lucy, the subject of Profile 86, cruises over the southern coast of New Guinea, just east of Port Moresby.

Profile 85   B-17E serial 41-2666 *Lucy*

This bomber first served with the 435th BS and became known as *Ole 666* when it was assigned to the 8th PRS on 27 July 1942. It was returned to the 435th BS in September, before it was again reassigned to the 8th PRS with which it operated for the first half of 1943. At this stage it was named *Lucy* in honour of Lucile Christmas, the daughter of Major General John K Christmas, an ordnance officer, whom one of the squadron's pilots had dated in the US. Reassigned to the 65th BS, it became famous as the bomber in which First Lieutenant Jay Zeamer was awarded the Medal of Honor, along with bombardier Second Lieutenant Joseph Sarnoski posthumously, after a mission over Bougainville on 16 June 1943. *Lucy* was returned to the US in March 1944.

Profile 86   B-17E serial 41-2627 *Lucy / Blanche / R.F.D. Tojo*

After previous service with the 19th BG and 64th BS, this Fortress was reassigned to the 8th PRS as a replacement when *Lucy* (see Profile 85 above) was taken offline to undergo repairs in June 1943. This bomber was also named *Lucy* when it was received in June 1942, however, this was soon painted over and changed to *Blanche* with an eight-ball motif. The bomber was referred to as *The Eight-Ball* on the flightline and the stork artwork of *R.F.D. Tojo* carrying a bomb was applied on the starboard nose. The bomber is illustrated with a replacement rudder and its retained anti-icing boot on the leading edge of the fin. On 26 December 1943 the bomber crashed on take-off from Schwimmer 'drome due to thick ground fog. Aboard were several reporters and two cameramen to observe and record the amphibious landing by US Marines at Cape Gloucester. The bomber was destroyed and several of those aboard were fatally wounded.

B-17E 41-2666 Lucy, as shown in Profile 85, seen after its June 1943 modifications including the addition of an extra six 0.50-inch calibre machine guns for defensive firepower.

B-17E 41-2426 from the 431st BS in HAD markings, preparing to return to the US in January 1944.

The artwork on B-17E 41-9227 Yankee Doodle Jr, the subject of Profile 90, at Henderson Field.

B-17F 41-24446 P'id Off Jr, as illustrated in Profile 88.

CHAPTER 20
431st Bombardment Squadron

Part of the 11th BG, this unit was based at Hickam Field during the Japanese attack against Pearl Harbor, when it was known as the 50th RS. On 16 January 1942 commander Lieutenant Colonel Walter Sweeney led six B-17Es on a special mission, three of which were from his 50th RS while the other three were from the 23rd BS. Their task was to assess the feasibility of conducting long-range ocean searches from advanced island bases. After returning to Hawaii the squadron was rebadged as the 431st BS on 22 April 1942.

Sweeney then led the 431st BS to Midway, where on 3 June 1942 nine Fortresses bombed a Japanese task force in three flights at different altitudes, but incurring no damage. From Midway the squadron's nine Fortresses redeployed to Nadi, Fiji, arriving on 24 July 1942.

On 1 August 1942 the 431st BS Advanced Echelon moved to Palikulo on Espiritu Santo, conducting attacks against Guadalcanal on 4 and 5 August. However, idle radio chatter between crews betrayed the approach of the Fortresses to the Solomons. Then, on 7 August 1942 the squadron suffered its most iconic loss. Major Marion Pharr, the squadron commander, departed Espiritu Santo early that morning at 0300 in B-17E 41-9220. Pharr's loss is sometimes explained via the possibility of friendly anti-aircraft fire, but no ship records indicate this. A clue lies when two carrier-based VT-8 TBF missions aborted in the face of bad weather over Malaita. It is more likely Pharr's bomber was the victim of this weather. He possibly descended through cloud to get a navigational fix, but instead flew into the ocean or Malaita itself, whose highest mountain reaches 4,275 feet. His Fortress remains MIA.

On night of 12-13 September three Fortresses, including two from the 431st BS ditched in bad weather when low on fuel, following a false sighting report of a Japanese task force. The crew from B-17E 41-2404 *Jitter Bug* took to a dinghy but two died before the remainder were rescued five days later. The second Fortress, 41-9226, had an engine disabled by anti-aircraft fire and was unable to keep up with the others. During the mid-afternoon return journey, it encountered an impenetrable wall of thunderstorms and by late afternoon the plane was far off course and unable to contact base. With just 40 minutes of fuel remaining, the crew sighted a tiny island which they circled for 30 minutes. Following a successful ditching, they were rescued by a PBY after four nights on the island.

On 1 November 1942 the 431st BS redeployed from Nadi to Palikulo before returning on 5 December 1942 to Nadi. This period included a deployment to Guadalcanal. On 7 January 1943 serial 41-2396 flown by Major James Edmundson ditched off San Cristobal with all of the crew later rescued. On 8 April 1943 the 431st BS departed Espiritu Santo to ferry their Fortresses back to Hickam Field where it would prepare for transition to the B-24 Liberator. The squadron lost six Fortresses during its SOPAC deployment.

The 431st BS logo was a simple obliquely split square, which derives from the squadron's original 1924 50th Aero Squadron design. It is shown at the top of page 108.

PACIFIC PROFILES

431st Bombardment Squadron

Profile 87 B-17E serial 41-2409 *Old Maid*

This bomber was painted in the HAD scheme following the attack on Pearl Harbor. It was one of the three 431st BS Fortresses which departed Hawaii on 16 January 1942 led by Lieutenant Colonel Walter Sweeney to survey the feasibility of conducting long-range ocean searches from advanced island bases. In his subsequent report, Sweeney stated that he considered the camouflage schemes on the HAD B-17s not particularly effective, especially as the red and white rudder stripes negated any camouflage advantage. The bomber is illustrated as it appeared at Nadi during January 1942. It was written off when it crash-landed at Henderson Field on 25 November 1942 after sustaining combat damage.

Profile 88 B-17F serial 41-24446 *P'id Off Jr*

This "F" model previously served with the 42nd BS (see Profile 36). It was transferred to the 431st BS on 15 September 1942 and was later used as a transport by Lieutenant General Nathan Twining. It was destroyed in a landing accident in Florida on 7 October 1944 after it had been returned to the US.

Profile 89 B-17E serial 41-2463 *Yankee Doodle*

This bomber departed Hawaii for Fiji on 15 November 1942 and then moved to Henderson Field on 3 January 1943. Whilst taking off for a sea search mission from Guadalcanal on 2 August 1943 the bomber failed to achieve take-off speed and ran off the end of the runway into palm tree stumps. It was destroyed by fire, and the bombardier and navigator were killed. The pilot was Second Lieutenant Eugene "Gene" Roddenberry, the future creator of the TV series *Star Trek*, who fortunately escaped injury.

Profile 90 B-17E serial 41-9227 *Yankee Doodle Jr.*

This Fortress was assigned to the 431st BS at Hawaii on 6 June 1942. It suffered a take-off accident departing from Palikulo on an intended New Year's Eve flight to Fiji on 31 December 1942, while flown by two inebriated officers intent on obtaining more alcohol for a party. Both men were killed in the accident.

Profile 91 B-17E serial 41-9128 *De-icer*

First Lieutenant Karl Stubblefield was flying this bomber on the night of 26 July 1943 over Buin (Kahili) airfield in clear weather. The Fortress was last seen five miles northwest of the target but was declared MIA when it failed to return. A statement in detailed Missing Air Crew Report 182 by Captain Roy Ballah is instructive:

> I was co-pilot in airplane [B-17E 41-2520] … over Kahili Airdrome … I saw our right wingman, Lt Stubblefield in airplane #9128 go down in flames. One or more enemy night fighters, unseen, attacked the formation and it was last seen to be in flames going down. Our own plane was also attacked and the assistant radio operator and tail gunner were wounded. Our aerial engineer saw one night fighter and fired at him. However, we returned to base safely …

B-17E 41-2667 Texas Tornado, as depicted in Profile 92, at Whenuapai near Auckland, New Zealand.

The wreckage of B-17E 41-2667 Texas Tornado after it crashed into a farm just after midnight on 9 June 1942 near Whenuapai.

CHAPTER 21
2nd Provisional Bombardment Squadron

In May 1942 the SOPAC air commander, Rear Admiral John "Slew" McCain, desired a detachment of long-range transports to ferry VIPs and senior officers between Pacific bases for liaison purposes. Such visits would enable his officers to oversee the training, indoctrination and operation of all SOPAC air units. The vast theatre contained territory which fell under five jurisdictions: USA, UK, Australia, New Zealand and France. Given the imminent invasion of Guadalcanal, the coordination of logistical, administrative and command matters required close collaboration by staff officers. With no dedicated air transport capability yet available in the theatre, one had to be created.

McCain's request was agreed by Hawaii to form a temporary detachment of five Fortresses for the job. Subsequently on morning of 16 May 1942, B-17 41-2658 flown by First Lieutenant Robert Irwin of the 42nd BS and 41-2667 flown by Captain Kenneth Bushnell of the 31st BS departed Hickam Field, bound for Tontouta, New Caledonia. Each B-17E carried four "specialists" in addition to their nine-man crews. Separately two LB-30s also departed carrying three more B-17 crews and sixteen additional "specialists". They delivered these crews to Nadi to collect three more B-17s to ferry to New Caledonia: 41-2617, 41-9015 and 41-2663. McCain's headquarters named the detachment the 2nd Provisional Bombardment Squadron, labelling it as a combat unit to obviscate its real purpose.

The detachment's first job was to deliver VIPs for multi-lateral conferences hosted in Auckland, the first set for 7 June 1942. This meeting would address the imminent arrival of American forces to New Zealand, as a springboard for the invasion of Guadalcanal. Follow-on conferences would take place on 15 and 28 June, with both attended by McCain. This second conference was presided over by the SOPAC commander, Vice Admiral Robert Ghormley, visiting from Noumea.

The detachment was initially based at Tontouta, however, the Fortresses tore up its fledgling dirt runway thus far capable of handling aircraft only up to C-47 size. On 1 July 1942 the 4,600-foot runway was closed for re-surfacing and lengthening, a task that took about two months. This explains why the first 11th BG B-17E detachments to New Caledonia arriving on 21 July 1942 based themselves at Plaine des Gaiacs and Koumac instead. During this phase the detachment's B-17s maintained operations between Australia, New Zealand and Fiji, leaving flights to or from Noumea to USN PBYs based there.

The 2nd Provisional Bombardment Squadron was disbanded around August 1942 with the inventory reassigned to combat units, mainly the 19th BG. This was in response to dedicated C-47 and R4D squadrons arriving in the theatre to take up the transport role. The squadron adopted no logo.

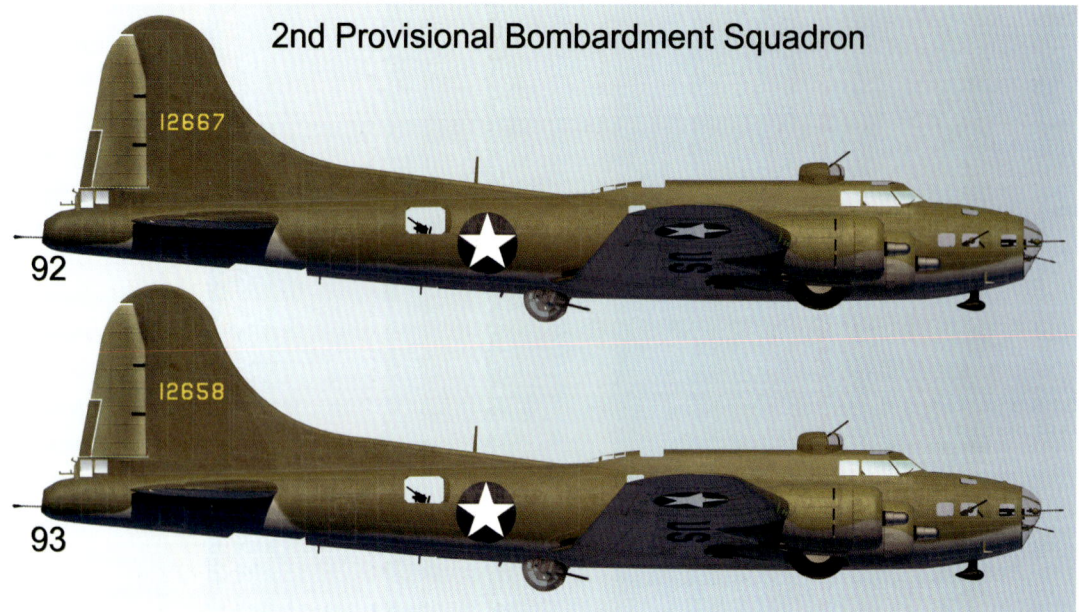

Profile 92   B-17E serial 41-2667 *Texas Tornado*

This bomber was ferried from Hickam Field to Tontouta in May 1942 with a 31st BS crew of nine and four extra engineers. After assignment to the 2nd Provisional Bombardment Squadron it arrived at Whenuapai, near Auckland, the day before the 7 June conference citied above. Just after midnight on 9 June 1942 *Texas Tornado* crashed on take-off from Whenuapai bound for Laverton near Melbourne. Just after lift-off pilot Colonel Richard Cobb suddenly banked into farm fields not far from the airfield. The bomber burst into flames and detonated two of the four 500-pound bombs aboard, shattering the wreckage into small pieces, and killing all eleven aboard. The cause was put down to either spatial disorientation, caused by failure to transition to instruments immediately after take-off, or leaving the control lock in place. It is unclear why the Fortress was carrying bombs, however, inaccurate coverage of this incident post-war sometimes delved into the hysterical. For example, one New Zealand newspaper in the 1980s claimed it was part of a "top secret" mission to exchange high-level Japanese prisoners. The bomber was allegedly named *Texas Tornado*, however, no photograph of the nose art has surfaced to date.

Profile 93   B-17E serial 41-2658

This bomber first served with 93rd BS and 435th BS in Australia. Shortly after transfer into the 2nd Provisional Bombardment Squadron in early June 1942 it was fitted with Automatic Flight Control Equipment for long-range flights which was tested at Plaine des Gaiacs. In August 1942 the bomber was returned to the 93rd BS at Mareeba, and it later became one of the bombers repatriated to Hawaii in October 1942. Following assignment to the 394th BS it was one of the Fortresses damaged by a VS-6 SBD on 25 November 1942 (see profile 75 for further details). As a result, it was written off.

B-17E 41-2633 Sally (not profiled), was the personal transport of Fifth Air Force Commander Lieutenant General George Kenney, at Seven-Mile in late 1943. The name underwent several iterations of calligraphy during its service.

Lieutenant General Walter Krueger (circled), the commander of the US Sixth Army, with the crew of B-17E 41-2462 Billy, as shown in Profile 97, at Townsville.

General MacArthur's VIP transport, XC-108 41-2593 Bataan, at Seven-Mile having its props turned to mitigate against hydraulic lock just prior to departure. This aircraft is the subject of Profile 98.

B-17E serial 41-2660 Muffins, as depicted in Profile 100, at Perth during its November 1943 circumnavigation of Australia.

CHAPTER 22
Commanders' Transports

Profile 94   B-17F serial 41-24381 *Well Goddam*

This Fortress participated in the last 63rd BS Fortress mission on 18 October 1943 (see Profile 43). The following month it was had its top turret removed at Garbutt and was reconfigured for service as a transport with the new nose art *Well Goddam*. It was later stripped to natural metal finish and converted into a VIP transport complete with blue leather seats added to the radio compartment. On 8 December 1943 it was reassigned for use as a transport by General Paul Wurtsmith, the commander of Fifth Fighter Command, and for the next two years it operated throughout Australia, New Guinea and the Philippines. At the end of the war, it was left at Okinawa where it was written off in a hurricane.

Profile 95   B-17F serial 41-24450

In mid-September 1942, the SOPAC USAAF commander, General Millard Harmon, and his staff were assigned two B-17Fs (41-24430 and 41-24450) for regional travel. This B-17F was the first "F" model to arrive in the SOPAC theatre, arriving in New Caledonia in late September 1942. It was used by Harmon until reassigned to the 72nd BS in early 1943. During their assignment to Harmon's staff, both Fortresses were directly under Thirteenth Air Force command, and do not appear to have been given an official detachment identity. After 41-24450 was returned to the 72nd BS, it was destroyed at Tontouta on 15 March 1943 when it ran off the runway and drove into parked TBF Avengers.

PACIFIC PROFILES

Commanders' Transports

Profile 96   B-17E serial 41-2644 *Los Lobos* "The Rover Boys"

This bomber served with the 28th, 30th and 93rd BS before being ferried back to Hawaii departing in October 1942. After overhaul, it was assigned to the 394th BS, and the future creator of the TV series *Star Trek*, Second Lieutenant Eugene "Gene" Roddenberry, flew most of his 89 missions as a co-pilot in this bomber. In August 1943 it was transferred to the Thirteenth Air Force Air Depot at Tontouta where it was converted to a transport for use by the depot commander, Lieutenant General George McCoy. It was finally scrapped at Tontouta in August 1945. The name *Los Lobos* and the wolf nose art were painted on the airframe at Espiritu Santo by Sergeant Charles Petrakos, with the additional name *The Rover Boys* added when it was refurbished at Tontouta.

Profile 97   B-17E serial 41-2462 *Billy*

Following damage from strafing at Seven-Mile in December 1942 when serving with the 93rd BS (Profile 69), this bomber was converted to a transport for the use of Lieutenant General Walter Krueger, the commander of the US Sixth Army.

Profile 98   XC-108 serial 41-2593 *Bataan*

This Boeing is exceptional insofar as it was converted in the Boeing factory to XC-108 VIP transport configuration in mid-1943. The aircraft's interior was converted into a flying lounge with extra windows, cooking facilities and living space added. It was ferried to Australia in October 1943, for use as General Douglas MacArthur's own transport. The name *Bataan* references the area where MacArthur's soldiers held out in the Philippines in 1942.

Profile 99   B-17E serial 41-2417 *Nancy*

This B-17 had a rich history (see Profile 83) before it was stripped down to natural metal finish and transferred to Fifth Bomber Command in July 1943 for use as an armed transport. In October 1944 it became the personal transport of Lieutenant General Clements McMullen, the commander of Fifth Air Force Service Command, who used it until the end of the war. McMullen named it *Nancy* after his wife.

Profile 100   B-17E serial 41-2660 *Muffins*

After first use with the 28th BS, in 1943 this Fortress was rebuilt as a transport and assigned to the charismatic Lieutenant Colonel Victor Bertrandias, commander of the 4th Air Depot in Townsville. Of Basque heritage, Bertrandias was a previous Vice President of the Douglas Aircraft Corporation, spoke Spanish and was a qualified command pilot. He named the aircraft *Muffins*, the nickname he called his wife. The ball and top turrets were removed, and extra wing tanks were installed to increase range. The rebuilt bomber's maiden flight was on 23 October 1943 when it was flown down to Brisbane to showcase it to Lieutenant General George Kenney. Shortly afterwards it completed a circumnavigation of Australia, purportedly to identify Australian bases to serve as frontline airfields in case of Japanese invasion. However, at this stage of the war the scenario was implausible and instead the trip enabled Bertrandias to visit many of his previous colleagues from his Douglas Aircraft days, then stationed all over Australia. *Muffins* was scrapped at Morotai in late 1944.

B-17F 41-24548 *Harry the Horse*, as illustrated in Profile 101, following its landing mishap at Tadji on 4 May 1944.

B-17F serial 41-24353 *Cap'n & the Kids*, the subject of Profile 104, taxies at Nadzab in 1944 when serving with the 69th TCS as squadron number 371.

CHAPTER 23
Armed Transports

In early November 1943 a mixture of a dozen "E" and "F" model B-17s were converted into armed transport configuration at the 4th Air Depot at Garbutt, before allocation to troop carrier squadrons. During the modification process several were given replacement "E" model Perspex noses and engine replacements. Inside the fuselage the ventral Bendix ball turrets were removed to reduce weight and make more room to carry supplies. The fuselage floors were strengthened, and bomb bay areas were modified to accommodate trays for air drop cargo. Not all modifications were uniform and several bombers underwent subsequent minor field modifications by their host units. The armament depended on each unit and often varied on the degree to which the mission was a "milk run". At least one was later converted to RB-17 status (see Profile 103).

B-17F serial 41-24353 in its final natural metal finish VIP transport configuration at Nadzab in 1945. It has been renamed Miss Em by Major General Robert Eichelberger after his wife. The bomber is illustrated in Profile 104 when serving previously as an armed transport named Cap'n & the Kids.

PACIFIC PROFILES

Armed Transports

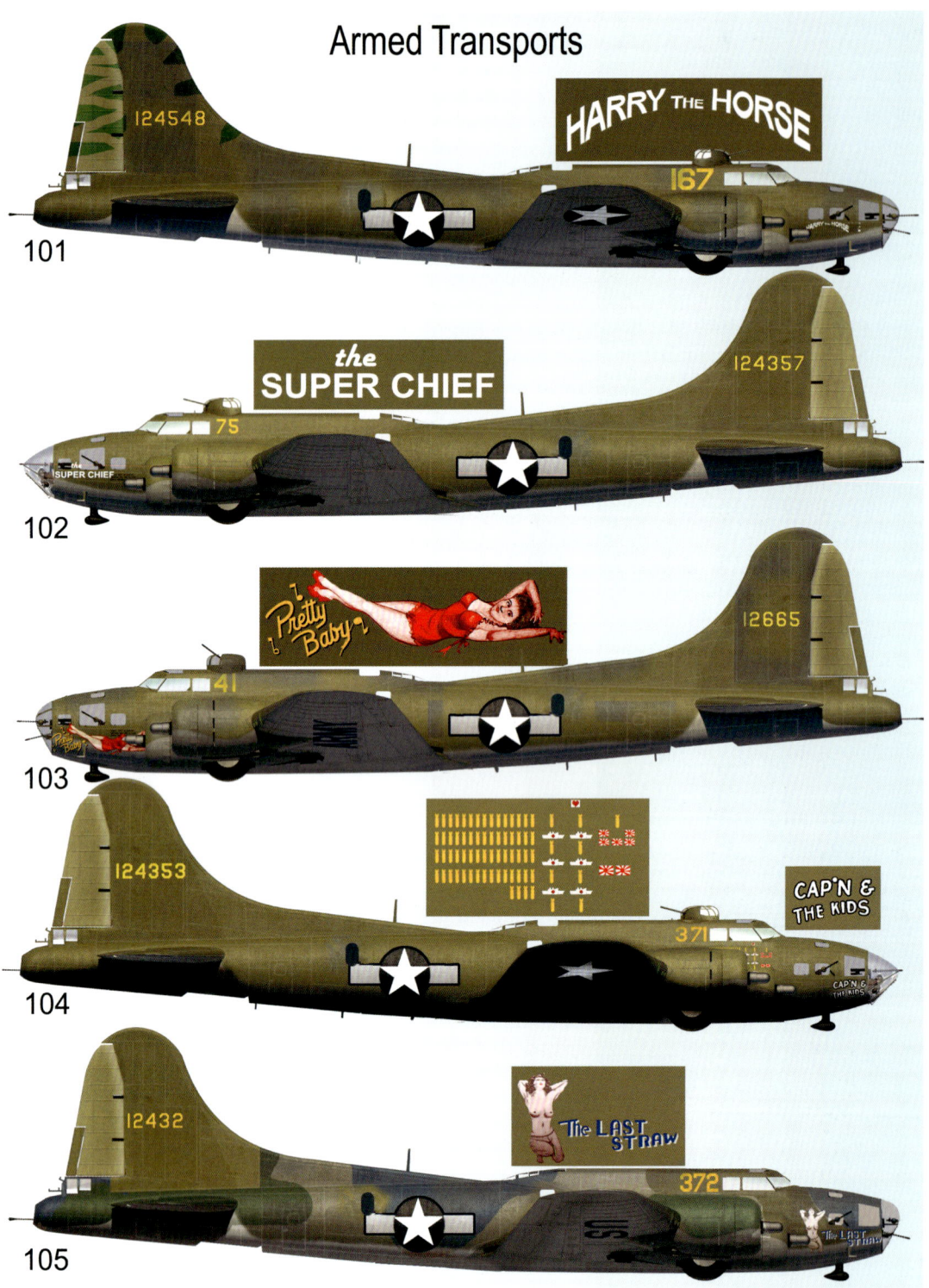

ARMED TRANSPORTS

Profile 101   B-17F serial 41-24548 *Harry the Horse*

This B-17F was first assigned to the 403rd BS in September 1942 where it was named *Little Poopsie Adele*. The following year it also served with the 64th and 65th BS. During conversion to an armed transport, it was uniquely given a replacement "E" model Perspex nose. After assignment to the 57th TCS it was renamed *Harry the Horse* with squadron number 167 painted in yellow on both sides of the fuselage behind the cockpit. On 4 May 1944 it departed Nadzab to drop supplies to troops at Hollandia. However, during the return flight it ran short of fuel in bad weather and diverted to Tadji, where it was written off in a landing accident.

Profile 102   B-17F serial 41-24357 *The Super Chief*

Delivered to the theatre on 6 August 1942, this "F" model first served the 63rd BS as *Tojo's Nite Mare / Moma Maxie*, and then with the 65th BS as *Blonde Bomber*. Following modification at the 4th Air Depot to an armed transport it was reassigned to the 41st TCS where it was renamed *The Super Chief* and was mainly flown First Lieutenant Jack Hoover.

Profile 103   B-17E / RB-17E serial 41-2665 *Pretty Baby*

This bomber was first assigned to the 93rd BS in August 1942 following which it served with the 64th BS (see Profile 52). After brief service with the 63rd BS it was converted to an armed transport in November 1943. It then served with the 40th TCS where it was renamed *Pretty Baby* and given squadron number 41. Later it was converted into a reconnaissance Fortress and redesignated as an RB-17. It then was based at Manila with the Far East Air Force, prior to transfer to the 13th Air Rescue Squadron with which it served until 1948. It was scrapped in the Philippines in January 1949.

Profile 104   B-17F serial 41-24353 *Cap'n & the Kids*

This bomber was named *Cap'n & the Kids* while serving with the 63rd BS. Note this was the only Fortress in the theatre with a matt black underneath, applied by the 63rd BS for night missions. It accumulated an impressive scoreboard of 80 missions during its combat service which was retained along with the name when it was converted to an armed transport. Assigned to the 69th TCS, it was allocated squadron number 371. In August 1944 it was stripped back to natural metal finish and reconfigured into a VIP transport for the use of Major General Robert Eichelberger who named it *Miss Em* after his wife. Its last operational flight occurred in the Philippines in August 1945, where it was scrapped the following year.

Profile 105   B-17E serial 41-2432 *The Last Straw*

In early 1942 this bomber served with the 14th RS (see Profile 3), and subsequently with the 63rd BS. Converted to armed transport configuration in November 1943, it was subsequently reassigned to the 69th TCS and given the squadron number 372. It was finally scrapped at Brisbane in January 1945.

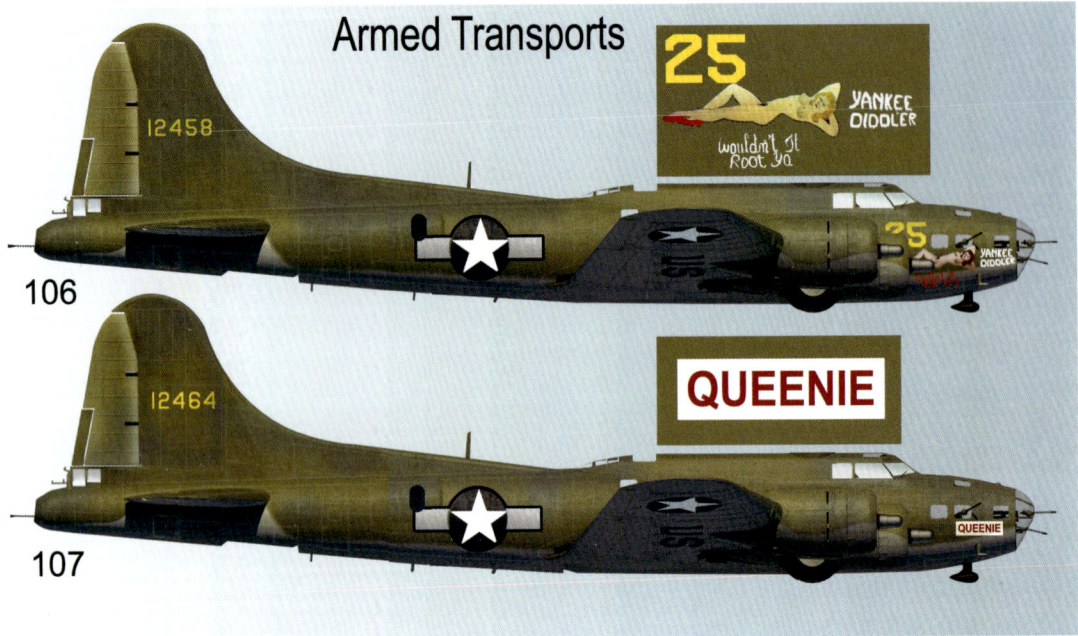

Profile 106   B-17E serial 41-2458 *Yankee Diddl'er*

This Fortress served in Java with the 28th BS and 30th BS and then from Mareeba with the 93rd BS and 435th BS. On 31 July 1942 it was reassigned again this time to the 8th PRS with which its ventral Bendix ball turret was removed, and the bomber was converted for photo reconnaissance flights. In November 1942 it was reassigned to the 65th BS then in February 1943 briefly to the 403rd BS, but by month's end it was returned to the 65th BS. Its original artwork was applied shortly after it arrived in Australia, but it changed in both colours and format over the course of the numerous units with which it served. When modified to armed transport configuration it was given four new engines, then assigned to the 39th TCS where it received squadron number 25. After an extensive career the bomber was finally scrapped at Brisbane in January 1945.

Profile 107   B-17E serial 41-2464 *Queenie*

This last profile in this volume is reserved for the last Fortress lost in the SWPA theatre. The bomber was a veteran of both the 19th and 43rd BGs. It was named *Queenie* (after a pilot's wife) when with the 64th BS, before it was transferred to the 65th BS in July 1943. On 3 August 1943 it suffered major damage when its copilot accidentally retracted the landing gear at Seven-Mile. After yet more repairs it briefly served with the Thirteenth Air Force in the Solomons as a photo platform. Then following conversion to armed transport configuration, it was transferred to Fifth Air Force Service Command. Its last flight was on 8 July 1944 when it departed Nadzab bound for Biak, but after take-off it was never seen again, costing the lives of five crew and fourteen passengers. The circumstances of its loss remain unknown, and the wreckage has not been located, giving strength to the possibility it may have crashed into the ocean.

Sources & Acknowledgments

This volume draws only on primary sources, with one exception. Over the past four decades fellow Australian Steve Birdsall has selflessly shared his knowledge on the B-17. His flawless research is second to none, and he qualifies as the foremost authority on Pacific B-17s. I accepted as gospel whatever information Steve provided; if unsure of a fact Steve will state so. The author's extensive collection of photos and notes from field trips 1964-2018 also contributed to the plethora of information herein, too extensive to list. Mainstream sources include, *inter alia;*

Pacific Aircraft Historical Society - Wreck Data Sheets

PNG Colonial Office - Civil Administration Records and PNG Cultural Museum

Aircraft Movement Entries, Townsville Control Tower, 1942-44

Field notes Robert Greinert, John Douglas and Brian Bennett PNG, 2011-2019

Papua New Guinea Catholic Mission Association, Field Trips, PNG

Website www.pacificwrecks.com and its diligent webmaster, Justin Taylan

Allied Translator and Interpreter Section (ATIS) Reports

Allied Air Force Intelligence Summaries (AWM)

ANGAU patrol officer reports of Allied crash sites 1940s-1970s

B-17 markings details within Individual Deceased Personnel Files (IDPF)

Microfilms/ official unit records

USAAF Fifth and Thirteenth AF Unit official histories via Maxwell AFB (including service sheets): Fifth Air Force Establishment, Fifth Bomber Command, Thirteenth Air Force HQ, 11th BG, 5th BG, 19th BG, 43rd BG, 14th RS, 23rd BS, 26th BS, 28th BS, 30th BS, 31st BS, 40th RS, 435th BS, 42nd BS, 63rd BS, 64th BS, 65th BS, 403rd BS, 72nd BS, 93rd BS, 98th BS, 394th BS, 8th PRS, 431st BS, 4th Air Depot and 27th Air Depot.

Boeing Deliveries and Departures from the U.S, By Model Jan 1942 to Dec 1943

War Diary Group Headquarters Fiji 1940-1943, National Archives New Zealand, AIR 129, Boxes 1, 2 & 3

USAAF individual aircraft record cards (AFHSO, Bolling AFB).

RAAF Iron Range, Cloncurry, Torrens Creek and Longreach airfields, aircraft movements

Index of Names

Allen, Colonel Brooke 19
Bacon, Master Sergeant Layton 65
Ballah, Captain Roy 109
Barnett, First Lieutenant Harold 63
Barr, Captain Bernie 97
Benn, Major William "Bill" 10, 63
Bertrandias, Lieutenant Colonel Victor 117
Blakey, Major George 29
Bleasedale, Major Jack 71
Bobzien, Lieutenant Colonel Edwin 19
Bostrom, First Lieutenant Frank 27
Brady, Major FT 47
Brecht, Major Harold 71
Brown, Vice Admiral Wilson 9, 25
Bushnell, Captain Kenneth 111
Carmichael, Colonel Richard 22, 25, 51
Charles, Major Thomas 99
Christmas, Major General John K 105
Christmas, Lucile 105
Cobb, Colonel Richard 112
Connally, Lieutenant Colonel James 22
Combs, Lieutenant Colonel Cecil 22
Cromer, Major Daniel 75
Dellinger, Major Russell 29
Earhart, Amelia 9
Eaton, First Lieutenant Fred 25, 27
Eberenz, Captain Richard 57
Eckley, Second Lieutenant Paul 101
Edmundson, Major James 107
Eichelberger, Major General Robert 119, 121
Everest, Colonel Frank 20
Ghormley, Vice Admiral Robert 111
Gibbs, Major David 21
Glober, Major George 47
Grant, Cary 101
Green, Major Franklyn 99
Guenther, First Lieutenant Robert 33
Hall, Major Earl 57
Halliwill, Captain Eugene 71
Halsey, Vice Admiral "Bull" 20
Hamlin, Vincent Trout 60
Hancock, First Lieutenant John 45

Hardison, Major Felix 85
Harmon, General Millard 115
Harp, First Lieutenant James 59
Hastings, Major Harold 71
Hawthorne, Colonel Harry 23, 75
Helton, Major Elbert "Butch" 39
Hobson, Lieutenant Colonel Kenneth 22
Hocutt, Major Ealon 71
Hoover, First Lieutenant Jack 121
Holguin, First Lieutenant Jose 77
Hirsh, Second Lieutenant Nathan "Joe" 67
Irwin, First Lieutenant Robert 111
Ivey, Captain William 93, 97
Joham, Captain James "Ted" 81
Kenney, Lieutenant General George 23, 117
Key, Captain Fred 41
Krueger, Lieutenant General Walter 85, 113, 117
Kudo Shigetoshi, FPO2c 63
Lee, Gypsy Rose 77
Lewis, Major William 51
Lindberg, Major Allen 71
Lucas, Captain Anthony 31
MacArthur, General Douglas 22, 53, 114, 117
Manierre, Major Ernest 57
McArthur, Captain Charles 73
McCain, Rear Admiral John "Slew" 111
McCoy, Lieutenant General George 117
McCullar, Captain Kenneth 65
McMullen, Lieutenant General Clements 117
Meehan, Colonel Arthur 19
Messerschmitt, Captain Kermit 59
Noonan, Fred 9
Norton, Lieutenant Charles 9
O'Brien, First Lieutenant William 67
O'Donnell, Major Emmett 21
Pease, Captain Harl Jr 22, 24, 53
Petrakos, Sergeant Charles 117
Pharr, Major Marion 107
Polifka, Lieutenant Karl 103
Quackenbush, Lieutenant Commander Robert 21

Ramey, Colonel Roger 22, 23
Richards, Captain Robert 57
Riddingstook, Major Don 79
Rigley, Major Orin 93
Roberts, Lieutenant Colonel John 23, 75
Roddenberry, Second Lieutenant Eugene 93, 109, 117
Rousek, Major Jay 99
Rudolph, Brigadier General Jacob 11, 12, 15
Salisbury, First Lieutenant Stanley 73
Sarnoski, Second Lieutenant Joseph 105
Saunders, Colonel Laverne "Blondie" 19, 20, 29, 34, 37, 91
Scott, Major Ed 63
Sewart, Major Allan 33, 34, 37
Smith, Captain William 75
Sogaard, First Lieutenant Folmer 67
Snoddy, First Lieutenant Richard 48
Steed, Major Thomas 43
Stubblefield, First Lieutenant Karl 109
Sweeney, Lieutenant Colonel Walter 107, 109
Turner, Captain John 92, 97
Twining, Lieutenant General Nathan 47, 59, 109
Unruh, Lieutenant Colonel Marion 19, 81
Vandal, Technical Sergeant Ernest 65, 101
Walker, Brigadier General Kenneth 22, 71, 77
Waskowitz, Frank 91
Welch, Major William 99
Whitehead, Lieutenant General Ennis 45
Williams, Captain Paul
Wisener, First Lieutenant Jack 65
Wood, Lieutenant Colonel Jack 43
Wurtsmith, General Paul 115
Zeamer, First Lieutenant Jay 105
Zumwalt, Major McLyle 93